C0-AVD-163

Against the Tide

Against the Tide

Whites in the Struggle Against Apartheid

Joshua N. Lazerson

Westview Press
BOULDER • SAN FRANCISCO • OXFORD

Mayibuye Books
UWC • BELLVILLE • SOUTH AFRICA

Published in 1994 in the United States of America by Westview Press, Inc., 5500 Central Avenue, Boulder, Colorado 80301-2877, and in the United Kingdom by Westview Press, 36 Lonsdale Road, Summertown, Oxford OX2 7EW

Published in 1994 in South Africa by Mayibuye Books, UWC, Private Bag X17, Bellville 7535, South Africa

This book is number 50 in the Mayibuye History and Literature Series. The Mayibuye Centre for History and Culture in South Africa is based at the University of the Western Cape. Focusing on all aspects of apartheid, resistance, social life and culture in South Africa, its aim is to help recover the rich heritage of all South Africans and to encourage cultural creativity and expression. The Mayibuye History and Literature Series is part of this project. The series editors are Barry Feinberg and André Odendaal.

Library of Congress Cataloging-in-Publication Data
Against the tide : Whites in the struggle against Apartheid
Joshua N. Lazerson.
p. cm.
Includes bibliographical references and index.
ISBN 0-8133-8487-7
1. Anti-apartheid movements—South Africa. 2. South Africa—Politics and government—1948- . I. Title.
DT1757.L39 1994
968.06—dc20
93-39180
CIP

ISBN 1-86808-193-1 (South Africa)

Printed and bound in the United States of America

The paper used in this publication meets the requirements of the American National Standard for Permanence of Paper for Printed Library Materials Z39.48-1984.

10 9 8 7 6 5 4 3 2 1

To Nancy,
for everything,
with love

and

To uShezi,
for the future

Contents

Preface xi
Acknowledgments xv

1 Introduction 1

Notes, 9

2 Radical White Activity in the Interwar Years 11

Roots of Radicalism, 11
Notes, 20

3 From Class to Nationalism: White Communist Party Members in the 1940s 23

White Communists and Party Leadership, 25
The Africanist Challenge, 27
The Communist Party and the National Question, 30
The Ambiguity of White Politicking in the 1940s, 33
Toward an Integrating Ideology: Multi-Racial Cooperation in the 1940s, 42
The National Convention Idea and the People's Assembly Campaign, 45
Notes, 48

4 Transition: From the Springbok Legion to the Congress of Democrats 53

The Springbok Legion Through the War, 54
The Springbok Legion, the Torch Commando, and Constitutionalism, 61
Defiance and the Formation of the Congress of Democrats, 64
Notes, 75

5 **White Radicals: A Collective Biographical Sketch** 81

The Jewish Presence, 82
Afrikaners and Other Heretics, 99
British Immigrants, 105
Notes, 111

6 **Organization and Intellectual Life in the Congress of Democrats** 115

Taking the Measure of COD, 117
COD's Educational Function and *The World We Live In*, 120
The "Revolutionary Intelligentsia" in the Early Alliance Period, 124
Notes, 134

7 **White "Democrats" and the Question of Identity** 139

Nadine Gordimer's *A World of Strangers*, 141
Living Non-Racialism in the 1950s, 145
The Complications of White Affluence, 155
Notes, 158

8 **COD, the Congress of the People, and the Freedom Charter** 161

COD, the Liberal Party, and the Congress of the People, 164
At the Congress of the People, 172
The Freedom Charter and Dissent Within the Congress Alliance, 175
Critiquing the Charter: Africanists and Other Skeptics, 181
After Congress: The Million Signatures Campaign, 187
The Treason Trial: Solidifying Congress Identity, 189
Notes, 192

9 **Building the United Front in the Late 1950s** 197

Broadening the Front: The Multi-Racial Conference, 198
COD and Internal Crisis, 201
COD and the Liberal Party: Building the United Front, 205
COD, the ANC, and the "One Congress" Question, 211
A Pan-African Postscript, 218
Notes, 220

10 White South Africans and Armed Struggle 225

Sharpeville and the State of Emergency, 226
Whites and the Emergence of the Armed Option, 230
The National Committee for Liberation, 235
Notes, 239

11 A 1990s' Epilogue: Non-Racialism in the Final Analysis? 243

White Activism in the 1970s, 246
The Revival of Mass Political Organization, 251
The Johannesburg Democratic Action Committee, the UDF, and the Cape Democrats: Rehashing the "One Congress" Question, 254
Non-Racialism, Transformation, and Power, 262
Skating on the Interregnum: 1990 and Beyond, 266
Notes, 268

List of Acronyms 271
References 273
Index 277
About the Book and Author 289

Preface

This book had its origins in my first trip to South Africa in 1983. During those three months, on the eve of the United Democratic Front's formation, I met white South Africans of different generations who were working -- as activists, journalists, lawyers, and in other capacities -- to end apartheid. Perhaps the most enlightening moments of the trip were spent with Helen Joseph at her home in a Johannesburg suburb, the same house to which she was confined as the first South African house arrestee in October 1962.

Helen Joseph was a founding member of the Congress of Democrats, the white partner organization of the African National Congress during the Congress Alliance period of the 1950s. The Congress Alliance was formed when organizations representing Africans, Indians, Coloureds, and anti-apartheid whites joined forces to protest the government's structure of systematic, legislated, racial- and ethnic-based oppression of the majority of South Africans. As a leader in both the Congress of Democrats and the non-racial Federation of South African Women, Joseph's anti-apartheid activities had spanned thirty years at the time I met her, and would continue for another decade, until her death at the age of eighty-seven in late 1992.

During one of my visits to Joseph's home that winter, the members of the Students' Representative Council at the University of the Witwatersrand came to visit Joseph (apparently an annual occurrence) and hear her recount the history, as she lived it, of multi-racial anti-apartheid struggle in the 1950s, including the presentation of the Freedom Charter at the Congress of the People, the women's marches on Parliament, and the Treason Trial, during which 156 members of the Congress affiliates were rounded up and charged with high treason. Listening to Joseph was a revelation, not so much for the general historical outlines of her story, as for her elucidation and incarnation of white South African involvement in the national liberation movement. From the remove of the United States, I had found the fact, and history, of white involvement difficult to grasp. Nor had I come across any treatment of this involvement at greater than incidental length prior to my experiences in South Africa.

Perhaps it was Joseph's grandmotherly appearance that magnified the impact of her own history as a white activist who had been banned, jailed, put under house arrest, and shot at. The experiences of Helen Joseph and other white South Africans raised broader questions in my mind: How many whites had been involved in anti-apartheid activities, over what period of time, and to what effect? What had led to these individuals' identification with the African majority's aspirations? Did they find general acceptance within the national liberation movement? What impact did whites make, as *whites*, on the organizational and/or ideological evolution of the national liberation movement? These are some of the questions that informed the researching and writing of this book. While I have never accepted the notion that any individual could write the absolutely definitive work on any historical subject, I feel particularly strong about that in this case. My hope for this work is that it will raise useful questions: about the nature of racial, national, and ethnic identifications; about the possibilities for, and factors limiting, alteration or transcendence of such identifications; and about the notion of a common society, and what one means, or implies, when one speaks of creating a non-racial society.

There are undoubtedly many different approaches and emphases one could bring to the material herein. I have no doubt there will be other treatments of this history in the years to come. Certainly, if any final word were possible, we would not be reading it in the near future. For the fact remains that the questions posed here remain very much alive in this interregnum spanning the demise of South Africa's known, charted, apartheid past and the post-apartheid South Africa-to-come.

A Note on Language

The word "black" today is commonly understood to be inclusive of Africans, Indians, and Coloureds. This usage grew out of the early black consciousness period of the late 1960s. In the late 1940s and 1950s, "black" generally referred to Africans alone, while "non-European" was used to convey the collective group of Africans, Indians, and Coloureds. In this work, the term "black" will be used to denote this collectivity, with the exception of those places where the term "non-European" appears in direct quotation.

The terms "multi-racial" and "non-racial" also present possible points of confusion. "Multi-racial" and "non-racial" were used more or less interchangeably throughout the 1950s to connote a complete lack of racial or ethnic barriers both physically and ideally. The Congress Alliance, composed of member African, Indian, Coloured, and white organizations, was considered a multi-racial collectivity because its member organizations looked to a South African future without racial or ethnic barriers. In the later 1950s some participants in the progressive organizations of the time began to draw a distinction between "multi-racial" and "non-racial" organizations. The Liberal

Party, whose membership was open to all South Africans, claimed in the late 1950s to be a *non-racial* organization, as opposed to the *multi-racial* Congress Alliance which maintained organizational affiliations along ethnic and racial lines. Today the term "non-racialism" reflects the absolute absence of racial or ethnic barriers. Therefore, today the Congress Alliance would not be considered a non-racial collectivity, because it was composed of four racially and ethnically distinct organizations. The African National Congress of the present day is non-racial because its membership is open to all without reference to ethnic or racial categorization. For the purpose of rationalization in this work, "multi-racial" will be used to connote organization along national lines and "non-racial" will be used to describe organizations with racially and ethnically diverse memberships. The terms will be maintained as they are found when in direct quotation.

Joshua N. Lazerson

Acknowledgments

The author of a work so long in gestation has many people to thank. At the outset, I want to thank Gay Seidman, who has been a friend and mentor since my undergraduate days at the University of California, Berkeley.

There are many people to thank for making my experience at Northwestern University stimulating and, generally, enjoyable. I particularly want to thank my advisor, Professor John Rowe, for his assistance and support. Professor Jim Campbell, who joined the department during my last year, was extremely helpful in honing the early drafts of this book, and his insights and excitement for the topic provided me an added energy boost just when needed. Professors John Hunwick and Ivor Wilks were both excellent instructors, encyclopedic references, and sources of encouragement, which I have much appreciated. Northwestern University's Africana Library was a key resource for this project, and I owe a great debt of gratitude to the professionals staffing it. Hans Panofsky, curator of the collection for twenty-eight years, was of great help and remains a friend, and I thank him on both counts. I also want to thank Dan Britz and Mette Shayne for their energy and aid above and beyond the call. Patrick Quinn, Northwestern's archivist, was also of great help in navigating the Carter and Karis collection of South African political materials. There are many friends and colleagues from Northwestern to thank for their support and intellectual stimulation up to the present, including Jean Allman, Hamidu Bobboyi, Keith Breckenridge, Lisa Brock, Cathy Burns, Nate Holt, Lisa January, Bob Kramer, Nancy Lawler, David Owusu-Ansah, Sita Ranchod-Nilsson, Carter Roeber, and Vicky Tashjian. I want to thank Professor Karen Hansen, in Northwestern's Department of Anthropology, for her support, advice, and friendship over the last seven years.

During research trips abroad I received help from many individuals and institutions. In England, I want to thank the staff at the Institute of Commonwealth Studies, and Dr. Shula Marks particularly, for their assistance. I also want to thank the library staff at the School of Oriental and African Studies. Dr. Stanley Trapido was kind enough to share his South African reminiscences with me and to extend the opportunity to present an early draft of this work at the southern Africa seminar in Queen Elizabeth House, Oxford

University. Dr. Harold Wolpe, then at Essex University and now at the University of the Western Cape, was also a source of aid and encouragement. David Everatt, who wrote a dissertation at Oxford and a number of papers, published and unpublished, on some of the same organizations discussed herein, was very generous in sharing both his insights and his collection of Springbok Legion materials from South Africa. I thank him for both. Many thanks also go to good friends Tim and Mique Cohen-Kuyck and Lisa and Neil Rose, all of London, for their friendship and support.

Ilse Mwanza, in the Institute of African Studies at the University of Zambia, was helpful and encouraging during my stay there. I thank both the Institute and the University of Zambia for extending the facilities of the university to me during my stay.

In Zimbabwe, the staff of the Popular History Trust in Harare were generous with both their time and the Trust's materials.

I was finally able to visit South Africa again in mid-1993 and have a number of people to thank there. The staff of the South African History Archives in Johannesburg were extremely helpful in providing me access to valuable materials at short notice. Kgomotso Jolobe was particularly helpful, and I thank her both for her aid and her willingness to share her insights on the last decade with me. In Cape Town, a number of people at the University of the Western Cape were greatly helpful in my short time there. I want to thank Dr. André Odendaal, director of the Mayibuye Center, for allowing me access to the Center's archives and for sharing his personal insights. Albert Fritz, the Center's archivist, was unfailingly helpful and enthusiastic, which I appreciated greatly. I also want to thank the University's History Department for the opportunity to present a portion of this work while there, and particular thanks go to Ciraj Rassool for his friendship and encouragement.

I am especially grateful to all those South Africans who were willing to take the time and care to share their recollections and interpretations of this period with me. They have afforded me the most fascinating, and often inspiring, experience of my life.

Professor Jean Allman, in the Department of History at the University of Missouri, Columbia, and Professor Sita Ranchod-Nilsson, in the Departments of Political Science and Women's Studies at the Iowa State University, have been particularly helpful in providing comments on draft material during the revision of the manuscript. Thanks also to Christopher Engholm for his valuable comments on the final chapter.

My editor at Westview, Barbara Ellington, made this a painless process and was helpful at every step in the preparation of the final manuscript. I thank her for her guidance and professionalism. I owe a truly special debt of gratitude to my mother, Arlyne Lazerson, an author, one of the best editors in the business, and a source of inspiration long before I conceived of this project. If this text is at all readable, it is largely due to her intelligence and keen eye. My love and gratitude go to my whole family: my father, Earl; my stepmother, Ann; my brother, Paul; my uncle, Daniel Steinberg; and my parents-in-law, Marge and Jay Lucker, for their years of support and encouragement and for understanding the importance of this work to me. I also wish to thank Richard Kinney, a long-standing publishing professional and good friend, for his advice and counsel.

Finally, I want to thank Rose Sleigh, my English teacher in my final high school year, for a year of patience and seventeen of friendship, and my other family, the Kimenyes of Nairobi, Kenya, for giving me a home in Africa.

J.N.L.

1

Introduction

Speaking at a 1986 Johannesburg conference on the role of whites in the "National Democratic Movement," Raymond Suttner, a law lecturer and an executive member of the United Democratic Front, suggested that, through their work with African, Indian, and Coloured organizations, whites could help "build the non-racial relationships and consolidate the unity which will be the basis of the South African nation of the future."[1] Suttner described the white democratic role in the nation-building process as "crucial." Suttner's remarks suggest, in declarative form, the questions that are the central organizing force of this study: How does one account for the rise of a non-racial political culture in South Africa? What role, if any, can be ascribed to white South Africans in the evolution of that culture? And does the microcosmic political history of white involvement in the national liberation movement shed light on broader questions, of the possibilities for nation-building in complex multi-racial and multi-ethnic societies, and the manner in which one understands the concept, "nationalism," itself?

The term "non-racialism" has come to connote the complete absence of barriers predicated upon race or ethnicity, whether in the context of an organization, a community or a society. Those organizations accepting of non-racialist doctrines have refused to abide by apartheid barriers and have both organized and socialized with members of all national groups, African, Coloured, Indian, and white. Understood as a process in which racial and ethnic barriers are progressively eroded, non-racialism has come to suggest, as alluded to by Suttner, the creation of a "new nation" that would define its citizens not in terms of racial or ethnic origin, but as South Africans.

Nowhere else in Africa have members of both the oppressed and oppressor groups joined to wrest state power from a white minority ruling class with a goal of redefining the nation without reference to racial or ethnic division. A cursory examination of the last thirty-five years of African nationalism, in the context of the African National Congress and its allies, suggests the persistence of this non-racial "tradition" anchored by the Freedom Charter, the lodestar of the Congress Alliance and its supporters. The Charter, acclaimed by 3,000 African, Indian, Coloured, and white anti-apartheid delegates at the Congress of the People in 1955, stated in its preamble that "South Africa belongs to all who live in it, black and white." When, in 1986, Raymond Suttner spoke of the role of white democrats, the ANC had upheld the Freedom Charter as its statement of principle and vision for over three decades. Suttner's words fall squarely within the context of the Charter.

This non-racial "tradition," however, has been far more problematic than such a cursory examination would suggest, and it remains so today. Perhaps some among those opposing apartheid have made particularly conscious attempts to unite because apartheid has worked so hard to divide. Still, if relationships among black and white South Africans represent the building blocks of a desired unity, such relationships have constantly been complicated by myriad factors. Most obvious among these factors is the nature of South African society, the progenitor of gross inequalities of wealth, rights, and power between its white and black communities. Inevitably, these inequalities have manifested themselves on the social plane, producing distrust and bitterness even among those committed to extreme sociopolitical change. Patterns of behavior, of command and deference, spawned by apartheid's divisions and inequalities, complicate relations across lines of color. Relationships forged between black and white in the political realm are highly charged, bearing the weight of this baggage spoken and unspoken, and remain sensitive to changes in the sociopolitical milieu.

This study is structured around three organizations in particular: the Communist Party of South Africa (CPSA) in the 1940s; the Springbok Legion, a veterans' organization, in the 1940s and early 1950s; and the South African Congress of Democrats (COD) from the early 1950s through the early 1960s. White South Africans who sought to support blacks in their efforts to change the status quo found expression for their energies in these organizations. However, it is crucial to understand that the crux of this study lies at a sub-organizational level, in the relationships drawn across lines of color and among the individuals peopling those relationships. This underlying focus on the more subjective dynamics of relationships and individuals is not meant to detract from the importance of organizational histories in this period. This study will illuminate those histories, aiding in the establishment of a much clearer picture of organizational action and interaction as such a picture emerges in the history writing of a post-apartheid South Africa. However, to focus solely on the organizational would endanger an examination of the no-less-crucial realm of personal interaction.

"Political" ideas as to the nature of South African society and the intersection of race and class were tested in the more immediate and subjective arena of personal relationships. These were charged relationships, given the different worlds whites and blacks came from, and brought with them. Whites brought their idealism, their guilt, and a desire to be recognized as decent human beings freed from association with a bankrupt system. Blacks brought their own visions, burdens of generations of anger and frustration, perhaps a generalized deference to whites and in some instances an appreciation, undoubtedly mixed with a measure of distrust and wonder, for whites who would willingly seek comradeship and a new identity across the divides of race and class.

Who were these whites, willing to divorce themselves from the apartheid orthodoxies of white South African society? What brought these whites to the African nationalist movement, and what, in turn, did they bring to that movement? The group of white "democrats" who comprise this study was numerically quite small throughout the period in question, never numbering more than 200 or 300. It was an eclectic group, though one formed from a number of identifiable sources. Many of these whites were Jewish, second- or third-generation refugees from the pogroms and upheavals in eastern Europe. Jewish South Africans brought with them memories of persecution, a familiarity with socialism, and in some instances working-class organizational experience. Another group had roots in the skilled mineworkers and industrial laborers who emigrated to South Africa from England, Scotland, Wales, and Ireland. The workers of the British Isles brought a sharp class consciousness and a long history of worker organization. Yet others came of poor Afrikaner families on the platteland, who, like their African counterparts, sent family members to the cities to supplement a dwindling agricultural income. The few Afrikaners who broke with their legacy and joined with other whites brought a sense of mission to the cultural mix. No one in the "democratic" camp had more to lose than Afrikaners like Bram Fischer and Bettie du Toit, who, by virtue of choices made, became the heretics of Afrikanerdom.

These whites brought a variety of skills and traits to their respective organizations as well. Among them were printers, writers, lawyers and journalists, fund-raisers and other skilled types. By virtue of their social-cum-racial position, whites could offer their black comrades an entrée to the material comforts and the tactically useful benefits of the white residential areas and of white affluence and mobility generally. Whites, with their generally superior formal education and fluency in the English language, coupled with their culturally derived willingness to lead and to teach, were often highly visible as speakers and lecturers.

Decisions on the part of black leadership as to the role of whites in the struggle were complicated by the extraordinarily complex and multi-layered nature of relationships between blacks and whites. In the urban world of 1950s South Africa, the opportunities for both working contact and personal contact between blacks and whites were extremely rare. This fact alone imbued such contacts with complication. And beyond questions of political practice were questions of profound personal transformation. Meetings of blacks and whites in an atmosphere of relative equality had the potential for inciting feelings of profound liberation and, obversely, were fraught with tension and ambiguity, laden as they were with possibilities for mutual misunderstanding and even hostility. These meetings across the racial divide were a source of validation for white "democrats." They eased whites' deep-seated anguish at the possibility of personal complicity in the apartheid system. For black South Africans who carried within themselves the baggage of oppression and an ingrained deference to whites, these linkages were potentially as liberating.

Inherent in the questions raised to this point is the assumption that, complexities, ambiguities, and hostilities aside, whites who sought a place in the politics of liberation found that place. Although non-racialism may seem to have proved a most durable doctrine, neither the doctrine, nor the presence of whites in the nationalist movement, has stood uncontested. Any attempt to understand the hostility produced by the white presence is complicated by the multiplicity of issues that that presence engendered. The very fact of these volunteers' whiteness, while the most obvious of factors, was clearly of great significance, of tremendous symbolic weight, and absolutely controversial in the minds of many African nationalists.

Less obvious a factor, but of equal significance in understanding the meaning and effect of the white presence, was the relationship of the "democratic" white fringe to ideologies of the left, particularly communism as in the Communist Party of South Africa (after 1950 the South African Communist Party or SACP). Throughout the apartheid period, the South African state railed against importers of foreign ideologies as the primary agents of unrest in South Africa. Communism has stood as a kind of symbol for all of South Africa's ills in the eyes of the state, propagated by white foreigners to sow unrest and foment revolution among the black masses. The state long said that communism's ultimate goal was the revolutionary overthrow of South African society and its replacement by a Soviet satellite regime.

Of all white South African "democrats," Joe Slovo, former head of the South African Communist Party, has been and, arguably, remains the individual most feared and despised by the majority of white South Africans. Slovo, as a white communist South African involved in liberation politics for over 40 years, represents all that is traitorous to supremacist whites. Slovo has worked tirelessly for the realization of African majority rule. As a communist, Slovo had come to symbolize the prospect of Soviet hegemony in the region (it was widely believed that Slovo was a colonel in the KGB) and the overthrow of capitalism at home. In Slovo, white supremacists found a challenger both to apartheid's rigid racial hierarchy, and to the system of economic exploitation upon which it was based. Consequently, the government and other prominent voices in South African society waged a near constant campaign through the period in question against communism and its supposed white masterminds.

Apartheid logic dictated that white communists had, by virtue of their superior intellects, to be the ringleaders of the Soviet initiative in South Africa: Africans were their innocent dupes.

Thus, by virtue of its invective, and backed by its coercive force, the South African state appeared unwittingly to define for the black majority the heroic status of the Communist Party as a defender of the oppressed majority against white tyranny. Or did it? Africans did not have to embrace communism to hate apartheid or work for its demise. In fact, there never was, nor has there ever been, a unity of opinion on the part of Africans and other blacks as to the efficacy of communism and socialism or their organizational incarnations. The variation in black response to leftist ideologies, including outright rejection, qualified if cautious acceptance, and wholehearted participation at the leadership level, can only be understood in the context of the history of white participation in leftist and liberation movements.

The issues revolving around this nexus are both diverse and complex. It is a fact that communism and other leftist ideologies were indeed imported by whites, who were the earliest proponents and interpreters of those ideas and organizers for them. As the "bringers of the Word," whites, who generally had superior formal educations, were naturally placed to act as teachers of theory and interpreters of doctrine in the context of the South African scene. In this white supremacist society, whites in the party or other organized political entities or alliances were accustomed to command, to speak out and debate issues of theory and practice. Because of their color, white radicals enjoyed greater access to resources, financial, professional, political, and otherwise. Thus, while the ideas communists and other leftists brought to black South African society might find a fertile audience there, a positive reception was not guaranteed. Black distrust of communism and communists could be fueled by the prominence of white party members, as whites and professionals and theorists, and by the foreign origins and grounding of the doctrine itself. The local party's intense interest in and attention to the Soviet Union and to events outside South Africa could provide cause for suspicion among people for whom communism and government by soviets paled in abstraction when set against the immediate realities of encroaching apartheid rule. Some blacks were convinced that supposedly well-meaning whites, radical or liberal sympathizers with African nationalism, were in fact plotting to take control of these African initiatives.

From the early 1940s, self-styled "Africanist" thinkers and organizers questioned the validity of interracial cooperation. Africanists believed that the struggle for African liberation would be sidetracked and possibly hijacked were Africans to work with Indians, Coloureds, and whites. Anton Lembede, who formulated the Africanist philosophy, believed that Africans would not win their freedom until they had achieved mental and spiritual liberation, by looking to their own history, heroes, and capabilities.

If Africans worked with "foreigners," those of other cultures and races, or if Africans took up foreign ideologies, particularly communism, they would never achieve that most necessary mental and spiritual liberation. They would never have faith in the efficacy of their own efforts. Some of the young Africans who subscribed to the Africanist philosophy in the 1940s later became proponents of non-racial collaboration and helped launch the multi-racial organizational efforts of the 1950s. Other young Africans, however, maintained their Africanist perspectives, and in 1958 Africanist proponents broke away from the ANC and formed a new organization, the Pan-Africanist Congress (PAC). The participation of whites, some of them communist, at all levels of the Congress Alliance, appears to have been one precipitating factor in the breakaway of the Africanists and the formation of the PAC. This fracturing in the national liberation movement, later evidenced in the rise of the black consciousness movement, continues to manifest itself in the current politics of liberation.

Disentangling the diatribe, rumor, and reality of white participation in the mid-century period is an important historical problem. One is able to find, on the one hand, blanket condemnations of the Communist Party and the white participants therein as Soviet dupes and cynical manipulators, and, on the other hand, hagiographic accounts of the party's leading anti-apartheid role and the selfless contributions of white communists therein. This study was undertaken to discover whether radical whites played an inordinately prominent role, or sought to impose a particular ideological point-of-view, on the African nationalist movement. Of equal importance are a number of related "why" questions: Why did so many of the whites who chose to involve themselves in anti-apartheid work become members of the Communist Party and/or Marxists? And why were other whites who were neither communists nor socialists throughout the period in question willing to work within or with organizations that caused those individuals to be identified as communists?

These questions are posed to probe the nature of whites who not only sympathized but identified with the claims of Africans and other blacks to full citizenship in a post-apartheid South Africa. In attempting to draw out that which was unique to radical whites, it will prove useful to contrast them with other white South Africans who were working for change in the mid-twentieth century. The comparison of white South Africans working in the liberal tradition with their radical counterparts facilitates discussion of the most elemental issues of upbringing and outlook that might explain tendencies to embrace either more gradualist or more revolutionary approaches.

At its most basic level, the liberal tradition has stood for the rule of law and the rights of individuals and private institutions in the face of centralized authority. The basic beliefs of modern liberals concern the individual's access to political rights, economic freedom, and the right of free speech. The liberal tradition in South Africa arose largely from the presence and works of the

London Missionary Society; many of the participants, white and African, in the joint reformist initiatives in the 1920s and 1930s were involved in, and/or were products of, Christian missionary initiatives. The term "liberal" as used to describe whites in the first half of the twentieth century was synonymous with "friend of the native."[2] White liberals met their black counterparts in a number of venues, including the European-Bantu conferences of the 1920s and 1930s, the South African Institute of Race Relations (SAIRR), founded by liberals in 1929, and the joint councils movement, which brought whites and African leaders together. After the abolition of the Cape African franchise in 1936, prominent liberals such as Edgar Brookes and J.D. Rheinallt Jones represented Africans in Parliament in those seats set aside for white "African representatives."

Liberals were moralists and pragmatists. They lacked a grand critique of South African society and had no extensive alternative program to offer their white counterparts, while holding out hope through the 1950s that the more moderate voice of the United Party might trump that of Afrikaner nationalism. Rather, liberals worked from a solid base of belief in right and rationality: Segregation and oppression of the African majority were both moral evils and absolute stupidities. Economically, the urban African presence was becoming progressively indispensable to the well-being of white South Africa, and that presence should be acknowledged and accommodated. Liberals' efforts in the 1930s and 1940s focused largely on the betterment of living and working conditions for Africans through social welfare work and Parliamentary advocacy. Using their positions as the elected representatives of qualified Cape Africans, liberals attacked pass laws and poll taxes and served on myriad government fact-finding commissions. Through these varied activities, liberals sought to bring the realities of African impoverishment and black oppression to the electorate. The limitations of liberalism, however, were manifold, including a general unwillingness to offend the powers responsible for segregatory initiatives and a vision that was internally incoherent or ill-formed, coupled with unconscious elements of bigotry that today would be considered absolutely racist.

Liberalism's next incarnation, the Liberal Party, was formed of liberal associations in the Transvaal, Cape, and Natal in 1953. The Liberal Party's formation came at a time of burgeoning apartheid legislation and the rise of African militancy in response to it. The Liberal Party would hew to an ambiguous line from its inception to 1960. During that period the Liberal Party was the only legal organization in South Africa whose membership was officially non-racial. The party had substantially larger numbers of African than white members at the height of its enrollment. Yet only in 1960 did it move to an acceptance of an unqualified universal franchise. As will be discussed in Chapter 8, one of the central debates raging among the Liberal Party leadership in the mid-1950s was the nature of the relationship the Liberal

Party should seek with the forces behind African nationalist initiatives, particularly with regard to the ANC and its Indian, Coloured, and white allies in the multi-racial Congress Alliance. It appears that the debate over this question within the Liberal Party was greatly influenced by how individual members perceived the ANC's white allies in the Congress of Democrats. A number of Liberal Party leaders believed fervently that the Congress of Democrats was a communist front organization that had largely succeeded in gaining control of the Congress Alliance and of the liberation initiative generally.

There is, perhaps, a degree of irony to be found in the Liberal Party of this period; its membership included a substantial percentage of African leaders yet it was unable to commit to unqualified African suffrage. It was rabidly anti-communist and convinced that whites were in control of African initiatives yet its own non-racial national committee was approximately 70% white between the years 1953-1968.[3] The nature of the Liberal Party membership, of the organization as a whole, and of its relationship to its more radical white counterparts offers a fertile field of contrast. Investigation of these contrasts may offer insights about whites who were either communists themselves or were willing to brave that labeling for the purpose of being identified with African programs and aspirations. While the Congress of Democrats and Liberal Party lessened their estrangement in the later 1950s, the mindset of their members seems to have been greatly different, and this difference is a key element of interest in this investigation.

At a time when the world's attention is focused on the renascent nationalism of eastern European states and sub-states, one might ask what lessons South Africans might take from that region's current strife, and, conversely, what South Africa offers to those pondering the relationship of race and ethnicity and region and class in the context of constructing, reconstructing, or destroying nations. The negative lessons of this historical period, most violently incarnate in the balkanized relics of Yugoslavia, are manifest: Personal and group identifications may be held in check for a time through the influence of grander ideological or geographical constructions, or through the agency of an overarching coercive force. However, these identifications, of clan, ethnic group, or race do not appear simply to fall away with time. They are not neutralized by modernity, technical sophistication, or global interconnectedness.

Might South Africa offer more positive lessons and a more optimistic potential outcome? Certainly, South Africa merits examination if one is posing broad questions about the intersection of race and class and ethnicity. As a state in which government's resources have been focused largely on creating and maintaining distinctions of race and ethnicity, including the capitalization of ersatz "ethnic homelands," any efforts joining South Africans across these divisions to defeat such divisive programs and supplant them with nation-maintaining distinctions of race and ethnicity, including the capitalization of ersatz "ethnic homelands," any efforts joining South Africans across these divisions to defeat such divisive programs and supplant them with nation-building ideas and programs are worthy of attention. The focus of this work, the nature of interactions among white and black South Africans working in concert to end apartheid, may offer a window, albeit microcosmic, on the complexities, difficulties, and possibilities inherent in creating, or attempting to create, unified nations from diverse peoples long schooled in the language of difference and unused to thinking in terms of commonalities.

Notes

1. Raymond Suttner, transcript of speech given at a conference on the role of white democrats, Johannesburg, January 1986, Popular History Trust, Harare.
2. Richard Elphick, "Mission Christianity and Interwar Liberalism," in Jeffrey Butler, Richard Elphick, and David Welsh, eds., *Democratic Liberalism in South Africa: Its History and Prospect* (Middletown, CT: Wesleyan Univ. Press, 1987), 66.
3. Douglas Irvine, "The Liberal Party, 1953-1968," in Jeffrey Butler, Richard Elphick, and David Welsh, eds., *Democratic Liberalism in South Africa: Its History and Prospect* (Middletown, CT: Wesleyan Univ. Press, 1987), 119.

2

Radical White Activity in the Interwar Years

Roots of Radicalism

The migration of English, Scottish, Welsh and Irish workers to South Africa following the mineral discoveries of the late nineteenth century helped inject socialist consciousness among the white South African working class. The immigrants included men like H.W. Sampson, a London compositor who arrived at Cape Town in 1892 and helped build the South African Typographical Union. Sampson was elected the first Chairman of the South African Labour Party (SALP) at the time of its inauguration in 1910.[1] Archie Crawford, a Glasgow fitter, published the *Voice of Labour* in South Africa between 1908 and 1912. Renowned socialists of the day wrote for the *Voice*, and Marx, Engels, Debs, De Leon and others were made accessible through its pages.[2]

While socialism was theoretically colorblind, many of its early proponents were racist, mirroring the attitudes generally held by whites at that time. Typically, socialists refused to acknowledge any struggle but the class struggle. They dismissed early black nationalism as bourgeois. Since they took for granted the primacy of the white worker at the head of the proletariat, these workers and organizers cut themselves off from the expanding black proletariat.[3] Organizers in the Cape were slightly more amenable to the idea of organizing their black co-workers. Robert Stewart, a Scot who arrived in 1901, formed the Social Democratic Federation and organized Coloured workers in the Cape. The fact of the matter remained, however, that, generally, white workers made overtures to their black counterparts only when it was in their interest to do so. In a number of instances African and other black workers downed tools at white request, but their sacrifices were quickly forgotten after settlement.[4] Through the first two decades of the twentieth century, one could point to only a handful of whites who believed in the organization of Africans.

Two men particularly were responsible for suggesting the importance of organizing the black proletariat: S.P. Bunting and David Ivon Jones. Bunting came to South Africa as a soldier during the Boer War, took a law degree and settled in Johannesburg. He was elected to the Labour Party Executive in 1912.[5] Jones, a Welshman, was orphaned at an early age and contracted tuberculosis. Seeking a healthier climate, Jones went to New Zealand and then to South Africa, where he settled in the Transvaal and took work as a clerk. Jones was elected general-secretary of the Labour Party in 1914.[6]

The declaration of World War I precipitated a crisis in the South African Labour Party. Both Bunting and Jones were opposed to South Africa's participation in the war. When the Labour Party officially took a pro-war stance, Bunting and Jones left the Party and formed the War on War League. Attempts to bring the renegades back to the Labour Party failed, and Bunting and Jones founded the International Socialist League (ISL).

Bunting and Jones's decision to break with the SALP was not related to the Labour Party's racial philosophy or policy: These renegades were not idealists or saints. Rather, in 1916 and 1917 the ISL suffered a number of electoral defeats at the hands of the Labour Party. These defeats suggested to the ISL's leadership that it was not going to win the white working class to revolutionary change. Thus, the ISL was faced with a choice. It could either compete unsuccessfully with the Labour Party for the votes of white workers or it could turn to the largely unorganized black population and create a mass base there. The ISL moved to organize Africans. Africans were invited to meetings. In 1917 the ISL held a public protest against the Native Administration Bill, one of the first instances in which white socialists spoke against legislation that did not affect them, or the white working class, directly. The ISL sponsored labor classes for Africans and helped form the Industrial Workers Union of Africa for African workers. African and white ISL members were arrested together in the Johannesburg bucket strike of 1918. These actions drew the ISL into closer cooperation with Africans and moved it toward a broader concept of struggle while further distancing it from the white labor movement.

Bunting and Jones had come to conceive of the socialist struggle as non-racial. It was therefore imperative that all workers realize their common interests against the mine, factory and farm owners. Yet the ISL continued to focus on class struggle as against national struggle. Bunting and company repudiated nascent African nationalism as racialist. The ISL leadership either refused to, or could not understand, the lay of the politico-economic landscape as seen from African eyes. From the African vantage one saw a solid bloc of white workers and landowners standing together, antagonists of the African, against enlightened British capital. The Communist Party of South Africa would ultimately provide the synthesis between the ISL's commitment to the

African worker as a worker and a commitment to African workers as members of an oppressed national group. This synthesis would allow radical working-class leaders to liaise with "bourgeois nationalists." However, before the party would turn to this conception it would learn a number of harsh lessons at the hands of the white working class.

The 1922 Miners Strike

In retrospect, the Communist Party's participation in the 1922 mine workers' strike and revolt was based upon false premises. Party members, most of whom were white, hoped that the strike would instill a generalized class consciousness in the workers, as the party understood the strike to be a fight against capitalist rule. The strike was actually a white workers' action, however, which was triggered by the Chamber of Mines' planned repudiation of an agreement pegging the ratio of white to black workers in the mines. The plan would have led to the firing of 2,000 white miners. A Council of Action was formed, led by the expelled Mine Workers Union members and working out of the CPSA headquarters. Bunting and other communists viewed the strike as a means of raising African wage standards by refusing to submit to the debasing of white wage standards. The majority of white miners were Afrikaner nationalists who formed "commandos," armed groups of miners, and these commandos (under the Council of Action) took control of the strike, melding Afrikaner nationalist ideology and worker demands. A Communist Party banner in Fordsburg declared, "Workers of the World, Unite and Fight for a White South Africa." Bunting saluted these Afrikaner commandos as "Red Guards of the Rand."[7] The communists thus stood, and were seen to stand, squarely behind the color bar. When white strikers turned against blacks in bloody and unprovoked violence, the party, with the government, spoke out against the victimization of Africans, and Africans as distant as Cape Town held protest meetings to bring attention to the white miners' violence. The eight-week strike ended in a two-week-long armed revolt, which the government quelled with all the force at its disposal.

Edward Roux, who participated in the events of 1922 as a member of the Communist Party, noted that it "would have sounded fantastic" were one to have advocated the unity of white and black workers at the time.[8] However, the few whites sensitive to the position of Africans could not achieve a lessening of tensions on the part of white workers toward their black counterparts. In 1924 Smuts was defeated, and a National Party--Labour Party "Pact" government was elected on a "civilized labour" platform. The Communist Party supported the Pact in the election. One result of the Pact victory was the 1925 Mines and Works Act, which codified the color bar in law.

The Communist Party's shift of focus from the white worker to the black majority might be traced to the Party's conference in late 1924. Bunting, reporting on the 4th Party Congress in Moscow, declared that an "all-negro liberation or anti-imperialist movement throughout the world may well be more potent for our common cause than anything our mere handful of white workers as such in South Africa can accomplish."[9] Bunting also suggested that the proselytizing of blacks must be executed by blacks themselves. The conference decided against applying for affiliation to the Labour Party as it had done in recent years, and Willie Kalk, a trade unionist, called on the party to "support every form of native movement which tends to undermine or weaken capitalism and imperialism." Kalk called upon the party to "fight for race equality of the natives."[10] The 1924 conference marked a push to Africanize the party. An African was elected to the party's Central Executive in 1925, and the Party paper began to publish articles in African languages in 1926. By 1928 1,600 of 1,750 party members were Africans.[11] Bunting, once a lone voice demanding that the party turn away from its narrow focus on white workers, had succeeded. However the party, now affiliated to the Communist International or Comintern, was about to be wracked by dissension as a result of the Comintern's unilateral declaration of the "Native Republic."

The "Native Republic" Thesis and Party Fracture

The Communist Party of South Africa was jolted by political and ideological shifts in the period between 1928 and 1935. Reckless changes in Comintern directives, divisions among members engendered by those changes, and the personal political machinations of a few party leaders nearly killed the party. Much of the trouble stemmed from a section of a draft resolution at the Comintern's 6th Congress referencing South Africa. The Comintern statement required the Communist Party of South Africa to promote the call for, "an independent native South African republic as a stage towards a workers and peasants republic with full equal rights for all races, black, coloured, and white."[12] The resolution, reflecting Lenin's 1920 thesis on the colonial question, emphasized the party's work among the peasantry and the poor white population. To some party members the notion of a "native republic" smacked of anti-whiteism and represented a foolish turning away from the proletariat. However, to understand the totality of damage done to the party it is necessary to follow the shifts in policy that flowed from the Comintern's dictum.

The origins of the Native Republic thesis lie actually with a member of the Communist Party of South Africa. A CPSA delegation composed of James La Guma, a Coloured Party member, J.T. Gumede, a member of the African National Congress (ANC), and Daniel Colraine, a white boilermaker from the Transvaal, attended a League Against Imperialism conference in Brussels in early 1927. There, the South Africans put forward a resolution calling for the "right of self-determination through the complete overthrow of capitalism and imperial domination."[13] The delegation then went to Moscow and held talks with Bukharin, head of the Comintern, and the Comintern's Negro Subcommittee of the Anglo-American Secretariat. The delegation's discussions with the Comintern appear to have led to the Comintern's draft South African thesis. The Comintern then sent the draft statement to the party in South Africa for discussion.

The majority of the party, led by S.P. Bunting, stood against the native republic slogan. The few in favor included La Guma, Douglas Wolton, the party's new secretary-editor, and the majority of Cape Town Party members. Bunting's criticisms of the Comintern draft had much to do with his belief that the Comintern members neither understood nor appreciated the particulars of the South African situation. Bunting argued that the Comintern talked about the "proletariat" of the industrialized nations but spoke only of the "masses" in colonial countries. He suggested that the Comintern assumed there was an African bourgeoisie when, in fact, it did not exist. He also noted that the ANC, virtually moribund at the time, had never demanded self-determination, but only equality. To Bunting, the class and national struggles of the African people were "practically coincident and simultaneous," and the slogan of a "native republic" would antagonize white workers and drive them toward the white bourgeoisie, closing off any possibilities of real worker unity among blacks and whites. Bunting wrote to Edward Roux, a party member who had originally accepted the Comintern draft:

> the language about "stages" represents the ideological rather than the chronological sequence (though I think it was dictated by the analogy of a bourgeois democratic native revolution in Cuba, but of course I didn't say that) as really no black republic in South Africa could be achieved without overthrowing capitalist rule. And I think the "stage part of the formula is verbiage....There is something not quite intelligible to the crowd about "Independent Native Republic." They all ask, "Well, if it doesn't mean driving the Whites into the sea, what does it mean?" and they don't want something that involves a lot of explanation.[14]

Bunting, his wife Rebecca, and Roux traveled to Moscow for the 6th Congress. Bunting attempted to amend the draft slogan to read "an independent workers' and peasants' South African republic with equal rights for all toilers irrespective of colour, as a basis for a native government."[15] The South African delegation was received with disdain, and the Congress refused to entertain the delegation's proposed amendments. Roux would later write that the Native Republic thesis had been forced on the Communist Party of South Africa by the Comintern. Roux in turn was criticized by Jack and Ray Simons, who insisted that the CPSA had "a strong tradition of internal democracy" and that the Comintern was not in the habit of imposing its will.[16]

The Simonses also suggested that Bunting and Roux, as "negrophilists," should have appreciated the "Native Republic" line, as it put the African majority firmly at the center of party work. This is a simplistic reading. The party had always stood for racial equality, and by virtue of population numbers, the expectation of governance by Africans was inherent in the party's understanding of the South African situation. In fact, Bunting did make peace with the Comintern directive and strove to put it into effect in 1929. The party was thriving at this time, and if it had been left to its own devices following the 6th Congress it might have maintained some of that momentum. Rather, the party found itself bombarded with a series of contradictory and confusing directives from Moscow over the next three years.

At first the party appeared to be making the best of a changed situation. The "Native Republic" directive was an invitation to return to United Front work and forge relationships with what were considered bourgeois black nationalist organizations. A free speech movement was started and included the ANC, the African People's Organization (APO), a Coloured political group, the Trades Union Council, the party and various African unions. Shortly thereafter a League of African Rights (LAR) was founded, headed by Bunting, Roux, Gumede, and Nzula. The League was ordered to engage in nonviolent protest aimed at winning specific civic and political rights, making it a reformist initiative in the classic Marxist sense. A petition against the pass laws was circulated throughout the country. Bunting, vying at the time for the Tembuland African representative seat, was able to spread the League's message quite successfully and enjoyed a great deal of support when he spoke, despite merciless hounding by the police. However in late 1929, the CPSA received a directive from the Comintern to the effect that the League was to be disbanded immediately and the party was to cease work with or in reformist unions. Martin Legassick has suggested that this leftward turn in party policy was a result of Stalin's attempts to rid the party of the left opposition. The Comintern chose to justify the policy shift as a reflection of crisis in the world capitalist system, which necessitated immediate preparation for the impending revolution.[17]

The Comintern's directive was followed shortly thereafter by the return of Douglas Wolton. Wolton, who had been in Moscow, set himself up as head of the South African Communist Party Central Committee. Wolton proceeded to purge many party members, including Bunting, Ben Weinbren, Bill Andrews and La Guma, describing them as "right wing, social democratic and vacillating elements."[18] Now the party's focus returned to union work. Wolton was seeking to politicize the unions by raising issues of unemployment and influx control. However Wolton, who had returned from Moscow with the purported purpose of "Bolshevizing" the party and bringing it back into the Comintern fold, now found that the Comintern's Executive was demanding that the party's union work revolve around wage demands, leading Wolton to the realization that in this erratic, Comintern-led dance, even his dictates were subject to overrule. Finally, in 1932, the Comintern abandoned the two-stage model of revolution, which had envisioned a national-democratic struggle and victory as a stage toward the realization of socialism. The "Native Republic" slogan was replaced with one calling for a "workers' and peasants' government" to nationalize mines and industries and redistribute land.[19] The party's policies had returned more or less to the point of origin, when S.P. Bunting had attempted to defend the national and workers' revolutions as coterminous. As the Simonses described it, Bunting's judges had "expropriated his policy without regard to the great debate on the two-stage revolution."[20]

Wolton and his wife Molly left South Africa in 1933. The estimated party membership at their departure was approximately 150, most of whom were white.[21] In four years the party had lost 90% of its membership. The party was thus quite weak when Georgi Dimitrov, new general-secretary of the Comintern, called upon communist parties to put all their energies toward the building of a people's front for the purpose of preventing war and stemming fascism at the International's 7th Congress in 1935. This new directive renewed antagonisms within the CPSA. Moses Kotane, John Gomas and others were in favor of the return to the United Front as a basis for party work. Lazar Bach, Wolton's lieutenant prior to his departure, continued to push the 6th Congress line. Bach preferred that the party not take a leading role in creating a United Front but work through organizations that might involve themselves in front work. Kotane, Gomas, and Roux contacted the Comintern and asked it to intervene in the dispute, which it did. Maurice Richter, a Free State Communist Party member, went to Moscow to argue the party political bureau faction's case, and was followed by Bach. Both were subsequently expelled from the party, put on trial and executed by the Soviet state.[22]

The second half of the 1930s was a fallow time for the party. The headquarters moved from Johannesburg to Cape Town in the hope that the change of venue would allow the Johannesburg group to straighten out its affairs. Members of the party were able to find some strength and calm in Cape Town. There, a group of party members, including Jack Simons, Ray Alexander (Ray Simons), Johnny La Guma, Moses Kotane, and Sam Kahn were able to pull the party back together. With a relatively strong core in Cape Town, the party was able to strengthen itself in the face of new political initiatives, particularly the formation of the National Liberation League and the Non-European United Front. Once again, the party began to look outward, to link up with the ANC, to organize black and white workers in unions. The party began, again, to move toward a broad alliance based in national struggle.

Martin Legassick, at the end of his paper on the "Native Republic" period, suggests that though the party was mending itself and moving forward from the late 1930s until the time of its banning in 1950, its theoretical understanding of nationalism was nevertheless impoverished. Communists did not understand that, in the South African context, struggles against aspects of the apartheid edifice such as passes were revolutionary struggles because the national struggle was the key to the destruction of the South African capitalist system upon which apartheid rested.[23]

A Counterpoint: Interwar Liberalism

If communists were groping toward a more sophisticated understanding of the nature of South African society and the prerequisites for revolution there in the interwar years, South African liberals found themselves even more divided and further from realizing their goals at the end of this period. If the Communist Party was wracked by intrigue and externally fueled differences of opinion regarding the nature of the revolution-to-come, liberal voices in South Africa proved incapable of responding either to African or Afrikaner nationalist initiatives; ultimately, they had little impact on either of these gathering forces.

Interwar liberalism was predicated largely on social welfare goals, as the state made no such provision for Africans. The evolving liberal critique of South African society in the 1920s, which identified the rise of a black professional stratum, was greatly concerned with the pressures created by the aspirations of such relatively well-educated and/or affluent blacks in that increasingly segregated society. The liberal program, lived largely through organizations such as the South African Institute of Race Relations (SAIRR), the Joint Councils movement, the social clubs for African men and a number of lecture clubs that brought white and African men together, provided forums in which black community leaders and white liberals could meet, black grievances could be aired, and white counsel and direction could be provided. The ANC's repudiation of such liberal initiatives isolated Africans willing to participate in them from the more progressive segments of African opinion.[24]

The key defect of liberalism, its inability to define an alternative vision for a South African society, led a substantial and influential segment of the liberal camp to acquiesce, more or less consciously, to the government's territorial segregation initiatives in the interwar years. From the 1920s a division in liberal thinking emerged between those more willing to accommodate to government policies -- for example SAIRR and its spokesman J.D. Rheinallt-Jones -- and a left-liberal segment whose most vocal exponents were William and Margaret Ballinger. The Ballingers were in turn linked to a coterie of liberals in London with particular concern for colonial issues and known as the Friends of Africa. Both liberal camps were influenced by the new anthropological work coming out of South Africa in the 1920s and 1930s.

The thinking of the dominant, more conservative liberal camp turned toward an acceptance of the government's segregationist activities; this turn was influenced by the apparent strength of the Afrikaner nationalist--Labour Party Pact government and the support expressed generally among white South African society for the Native (Urban Areas) Act and the Hertzog bills. These legislative initiatives were the bulwarks of African segregation and marginalization in both urban and rural areas, banishing Africans from the common roll in the Cape and robbing Africans of basic rights of residence and ownership outside of narrowly delimited areas in the cities and rural areas. As liberals lost influence in government decision-making processes, mainstream liberal initiatives focused more closely on welfare work, deserting the political arena.[25] With legislated urban segregation accomplished and government policy about the rights of Africans focused on the Reserves, mainstream liberal thinking in the late 1920s and 1930s accommodated itself to this new reality. The liberal majority continued to fight for retention of the Cape franchise, because it believed it had to coopt the African petit bourgeoisie as a buffer class. But most liberals acquiesced to the government's vision, which now defined the African as a rural dweller under the control, albeit modified for bureaucratic and coercive purposes, of a chief or other "traditional" authority figure as exemplified by Lugard's indirect rule.

The left-liberal reaction to the government's implementation of segregatory legislation was to turn an eye from South Africa to the British protectorates in the hopes of staving off those territories' incorporation and of facilitating African economic independence through the development of cooperatives, which might also serve to detract from the power of chiefs.[26] The Ballingers believed cooperatives could provide the link between "tribal" and market cultures and so could gradually integrate Africans into western economic and cultural modes. While the Ballingers' anthropological observations bolstered their perceptions that traditional authority must be devalued and African economic independence increased, the general South African anthropological trend in the 1920s and 1930s emphasized the persistence of African cultures in the face of the diffusion of western civilization and the differences between cultures. The recognition afforded African culture by South African anthropologists influenced the liberals working in SAIRR and made them increasingly amenable to segregation as a vehicle both for protecting African cultures and for managing race relations. The majority of Africans would live and farm in the newly expanded reserves, and a small urbanized class of Africans would live permanently around the cities. Or so went the mainstream liberal vision.

By the end of the 1930s the accommodationist liberalism centered in SAIRR prevailed. Rheinallt-Jones and Edgar Brookes were elected as Native Representatives in 1937, the initiation of their political tenure paralleling the promulgation of the Hertzog bills, with the purpose of socially, culturally, and economically isolating Africans. As Africans in the ANC and All African Convention (AAC) lost whatever faith they had held in liberalism as a vehicle for positive social and political change, the bases of apartheid were everywhere in evidence, although the decisive African nationalism of Anton Lembede and the ANC's Youth League was yet to surface. Where was the Communist Party as the basic tenets of apartheid were unfolding?

One might describe the CPSA in the 1920s and 1930s as having been lost in a theoretical wilderness. Rigorously in synch with the doctrinal gyrations of the Comintern, the CPSA leadership attempted to mediate between its own observations of the South African scene and the proclamations received from Moscow which were based less upon any intimate knowledge of South Africa than generalized distillations of colonial experience and palace intrigues. If the shape and content of Afrikaner nationalism in the context of a burgeoning South African industrial economy was visible in this period, an independent and forceful African response was not yet in evidence.

Party doctrine by the later 1930s took for granted the ultimate realization of a *non-racial* government composed of a majority of African workers and peasants. But there was a certain aridity to this vision, largely because it was based more on theory than on practical experience. Ironically, whites and blacks were living non-racialism through their interactions in the CPSA, but the full meaning or weight of that aspect of the party's work and experience was subordinate to discussions of theoretical issues and concerns focusing on the white working class.

It is very difficult to know what the non-racialism of the CPSA in that period meant, on a personal or local level, for those participants, black and white, who worked together to build it. One has a sense however, if one sets the CP of this period against the party of the later 1940s and beyond, that the party had yet to become truly three dimensional; it was not yet grounded in South African realities or fully cognizant of its primary constituency, the African majority. It was a party of workers but could not agree as to the nature and composition of the South African working class. It was a party that, at times, enjoyed the support of a majority African membership yet was oriented to a revolution and a leadership in Europe, and was largely led by whites who wrote and spoke and were steeped in the language of that revolution and its theoretical antecedents. It was a non-racial party flying in the face of South African convention, yet seemingly without a strong consciousness of the revolutionary nature of its non-racial experiment.

Notes

1. Jack Simons and Ray Simons, *Class and Colour in South Africa, 1850-1950* (London: International Defence and Aid Fund for Southern Africa, 1983), 107.

2. Simons and Simons, *Class and Colour*, 141.

3. This analysis is not uncontested. Baruch Hirson claims that these early white workers and organizers could not find an African proletariat, and could therefore not organize what did not yet appear to exist. Baruch Hirson, paper presented at the African Studies Association meeting, Atlanta, November 1989.

4. The Kleinfontein strike is discussed in Simons and Simons, *Class and Colour*, 156-160, and Edward Roux, *Time Longer Than Rope: The Black Man's Struggle for*

Freedom in South Africa (Madison: University of Wisconsin Press, 1948), 146-147.
5. Roux, *Time Longer Than Rope*, 130.
6. Simons and Simons, *Class and Colour*, 263-268.
7. Ibid, 286.
8. Roux, *Time Longer than Rope*, 150.
9. Brian Bunting, *Moses Kotane: South African Revolutionary* (London: Inkululeko Publications, 1975), 27.
10. Ibid., 28.
11. Ibid., 29-30.
12. Baruch Hirson, "Bukharin, Bunting and the *Native Republic* Slogan," *Searchlight South Africa*, Vol. 1, No. 3, (July 1989): 61.
13. Martin Legassick, "Class and Nationalism in South African Protest: The South African Communist Party and the 'Native Republic' 1928-1934," *Eastern African Studies* XV (1973), 11, published by the Program of Eastern African Studies, Syracuse University, Syracuse, New York.
14. Correspondence, S.P. Bunting to E. Roux, 5 December 1928, quoted in Baruch Hirson, "Bukharin, Bunting and the 'Native Republic' Slogan," *Searchlight South Africa*, Vol. 1, No. 3, (July 1989).
15. Brian Bunting, *Moses Kotane*, 37.
16. Simons and Simons, *Class and Colour*, 405-406.
17. Legassick, "Class and Nationalism," 20.
18. Roux, *Time Longer Than Rope*, 256.
19. Legassick, "Class and Nationalism," 51-52.
20. Simons and Simons, *Class and Colour*, 459.
21. Roux, *Time Longer Than Rope*, 269.
22. Simons and Simons, *Class and Colour*, 477.
23. Legassick, "Class and Nationalism," 67.
24. Paul B. Rich, *White Power and the Liberal Conscience* (Manchester: Manchester University Press, 1984), 24.
25. Ibid., 27.
26. Ibid., 40.

3

From Class to Nationalism: White Communist Party Members in the 1940s

Throughout the decade of the 1940s, the Communist Party of South Africa (CPSA) remained, at the leadership level, inordinately a white party. White Communist Party members would play a significant role in guiding the party through a decade that witnessed the fruition of Afrikaner nationalism, the genesis of African nationalism, and the initiation of pan-ethnic, multi-racial social and political protest. The party and its leadership were dramatically challenged to respond to the sweeping changes of the decade.

Entering the 1940s, in its analysis of the South African scene, the party was strongly materialist. While the party acknowledged and worked with the national organizations of the African, Indian, and Coloured communities, party leadership persisted in the belief that a vanguard of workers, cutting across racial lines but led by the white working class (as the most advanced portion thereof), would lead a socialist revolution in South Africa. The events of the 1940s shook the Party's analytical presumptions, propelling it toward a more favorable conception of national struggle. Yet one would be hard pressed to make a case that by the time of its banning in 1950 the Communist Party had abandoned its belief in the primacy of the working class. The question of the relationship of class and national struggle in the apparently unique conditions of South Africa, the "national question," weighed heavily on party theorists and policy makers throughout the 1940s. While there was no resolution of the national question for the party by the time of its banning, the realities of national struggle impinged on both party work and doctrine. There was a shift in emphasis from class to nationally based or national liberatory activity, and this shift raised particularly serious questions, both theoretical and personal, for white party members.

White Communist Party members had been greatly influential in setting the analytical parameters of party debate from the party's inception, and they remained so throughout the 1940s. During that decade the party Central Committee's membership was never less than 50% white. White party members found a rational basis for their self-identification as leaders in the nature of party doctrine.

The workers' struggle was one of class, not race, which was understood as an obfuscation used by the ruling class to hide the character of generalized class exploitation. The theoretical and philosophical underpinnings of the party made the central role of whites an explicable, even natural, one. These individuals, as members of the privileged white caste, were generally better educated and more conversant in political theory than their black counterparts. The fact that they were members of a non-racial party, a Marxist-Leninist vanguard party, theoretically colorblind and attuned to questions of class, probably mitigated any concerns white party members may have had as to the prominent role they were playing in a situation in which they were, objectively speaking, members of the dominant, oppressing caste.

Nationalism, particularly African nationalism, thus threw up two particular challenges to white Communist Party members in the decade. White communists were challenged as orthodox Marxists who had looked to the working class to lead a workers' and peasants' revolution, itself colorblind. In the 1940s it became clear that the white working class was lost to the party, and lost to the concept of worker unity itself. Party members were highly reluctant to abandon their belief in the working class, though it was not immediately clear who would now challenge the state and how they were going to do it. Theory would have to conform to a changed reality.[1]

The second challenge to white party members, and a challenge of a more personal nature, raised a question of place. White party members had had little call for publicly justifying their participation and leadership in the Communist Party of South Africa. One could surmise that for most Communist Party members, primary organizational allegiance was to the party, other organizational commitments were peripheral to that of the party, and one's understanding of and commitment to other political forms and organizations moved outward from one's Communist Party experience.

African nationalism offered an alternative focal point for political energies and aspirations. And this nationalism, antagonistic to communism throughout the 1940s, was completely and unabashedly conscious of race and of history, sought no concord with white sympathizers, and offered no home for them. Whites in the Communist Party had to ask themselves not only how the Communist Party might articulate with the rising, vibrant nationalism of the 1940s, but whether the white members of the party would have roles to play on this changing political and ideological landscape.

White Communists and Party Leadership

There are few estimates of general party membership for the 1940s. The figures that pertain to leadership positions, such as the composition of the party's Central Committee or district committees are well-documented and reliable. One is struck by the role of whites in party leadership positions. In 1944 whites composed 64.3% of the Central Committee; the number declined to 58.8% in 1950.

The two focal points of party activity were Johannesburg and Cape Town. The 1947 Cape District annual conference hosted 54 delegates, who claimed to represent 700 party members.[2] The 54 delegates elected a District Committee of 12, with Harry Snitcher, a long-time party activist, as chairman and Fred Carneson, secretary. The other Committee members were W. Forgus, I.O. Horvitch, A. Ludski, Bertie Louw, Hettie McLeod, S. Mfaxa, J. Morley, Johnson Ngwevela, Jack Simons, and Sara Carneson.[3] The roll call of the Johannesburg District Committee exhibited the same preponderance of whites. Yusuf Dadoo, the party stalwart and a highly respected doctor, was elected chairman, and Danie du Plessis, an Afrikaner, was elected secretary. Other members included Michael Harmel, Rusty Bernstein, S. Magomotsi, Edwin Mofutsanyana, Josie Palmer (also known as Josie Mpama), Issy Wolfson, Hilda Watts, the veteran African communist J.B. Marks, W. Roberts, Ruth First, and Joe Slovo.[4] White communists who were to take leadership roles in both the party and nationalist organizations of the 1950s and well beyond, including Joe Slovo and Ruth First, Hilda Watts and Rusty Bernstein, gained leadership experience through responsible party positions in the 1940s.

While the number of whites participating at district or national leadership levels is notable, it is interesting also to note the youth of the Committee members elected in 1947. Edwin Mofutsanyana, born in 1899, was the oldest African member of the Committee at 48 years of age. No date of birth is available for Palmer, also African, but she joined the party in the 1920s, so one could surmise that by 1947 she was in her later 30s. J.B. Marks, born in 1903, was 44, and the Indian leader and communist Yusuf Dadoo was 38. No birthdates are available for either Du Plessis or Wolfson. Of the other five white members, Michael Harmel and Hilda Watts were the oldest, born in 1915 and so 32 years old in 1947. Bernstein, born in 1920, was 27 at the time, and Ruth First and Joe Slovo, born in 1925 and 1926 respectively, were 21 and 22 years of age.[5] Why was there such a disproportionate number of whites in leadership positions, and why were these leaders so young relative to their black communist peers?

Consciously or unconsciously, white Communist Party members were the subjects of a fundamental contradiction. Notwithstanding their avowals to destroy the society they repudiated, it was a society in whose mores, processes, and values they were steeped. The realities of white supremacist government dictated that, by virtue of the color of their skins, they would enjoy the benefits accruing to full citizens of the dominating culture and its institutions of education, finance, and adjudication. Being white in South Africa, having as one's primary language the language of trade and law and education, put one at an obvious advantage, as did having access to capital, to secure lodging and automobiles and telephones. A less tangible but equally important consequence of being a white South African, and a consequence that demands examination in the relationship of white and black, within the party or without, was the acculturation to leadership, white over black, that was part of every white South African's experience. It is this element of acculturation that helps explain why whites, even twenty-year-olds, brought a certain innate self-confidence to their party work. The opposite side of this coin, initially if not in the longer term, was a built-in deference to their white comrades on the part of black party members. It was this sense of psychological inferiority, of self-bondage and deference, that the Africanists, proponents of African nationalism and founders of the ANC's Youth League, believed could only be eliminated through a separatist philosophy and a "go-it-alone" program. Party members, both black and white, were conscious of this reality. From time to time it was discussed in party media:

> Our more advanced comrades, who are usually white because of the superior social advantages of the European section, should step back where necessary, to allow non-Europeans to take the lead. In our committees and groups, do not let the position develop where one or two comrades monopolise the discussion, examine the subject under discussion from every point of view, analyse all the aspects -- and leave nothing for others to say! This practise deprives the less experienced comrades, who have not such ready tongues, of the chance to show initiative and develop judgment.[6]

Moses Kotane, one of the most prominent members of the party, wrote in 1942 that "[I]n a party like ours, where whites and blacks come together, the general tendency of Non-European members is to take back seats and leave the leadership to the Europeans. They feel themselves inferior to the European comrades. The reason for this is to be found in the political, economic and social structure of South Africa."[7] Perhaps there was a tension within the party between immediate leadership needs and the "grooming" of younger or less educated or experienced members. One might find such a tension in any organization seeking to balance present and future leadership. The Communist Party's dilemma in this context was distinguished by the fact that the gulf in skills and self-possession visible between white and black members mirrored the conditions of life in the nation generally. The party was dedicated to the ending of these injustices, and the party's non-racialism was a given. Yet, did party members ever raise the possibility of consciously de-emphasizing the white role in a party that was seeking to influence, to mold, a black struggle?

The answer appears to be no. The party's non-racial doctrine was not a guarantee that a numerical parity of races would be enforced at the leadership level. It could be argued, in fact, that the party's non-racialism, coupled with its self-defined vanguard role, worked to de-emphasize issues of race and ethnicity in party affairs.

The Africanist Challenge

Communists surveying the state of African organization in the early 1940s might have felt smug in their certainty that the African National Congress (ANC) was petit bourgeois and bankrupt. Since the 1920s, Africans had found themselves dispossessed in their own country, stripped of land, taxed, regulated in nearly every aspect of their existence. The old-guard African leadership had counseled patience while attempting to prove through word and deed that Africans could civilize themselves in the western mold. Looking to white spiritual, intellectual, and financial benefactors for confirmation and counsel, the leaders in the African community who gave birth to the ANC and saw it through its early years hoped that the British liberal tradition would prevail over Afrikaner nationalist zealotry. In a more general sense, Africans of all classes learned that feigned subservience was a survival skill in a society that had rendered them largely powerless. African ancestors of the nineteenth century had fought vigorously, if without an all-encompassing "African" consciousness, against white intrusion. In the twentieth century, as the consciousness of common oppression transcending ethnic boundaries grew among Africans, so too did the realization that Africans had little leverage in the face of white initiative, on the street or in Parliament. African deference to whites was the norm, and this bred in turn a psychology of hopelessness and an occlusion of African belief and feeling behind a servile, self-demeaning, and non-threatening facade.

But if the ANC entered the 1940s stagnant under the leadership of the liberal old guard, it would leave that decade under new leadership, ideologically revitalized, and with a program of action. The primary moving force in this transformation was Anton Muziwake Lembede, considered the father of Africanism, a philosophy that spoke directly to the psychological crisis of the African people as Lembede understood it.

Lembede was born into poverty in rural Natal in 1914.[8] Though poor, his parents put great faith in education. Lembede's mother had taught school for some years prior to marriage and provided her son early instruction at home. A family move offered the young Lembede the opportunity to attend a Catholic mission primary school. Lembede remained a practicing Catholic throughout his life. Upon completion of these studies Lembede entered Adams College to train as a teacher. A rather rarefied liberal atmosphere prevailed at Adams. A sharp and driven intellect, Lembede completed both the normal school and, on his own time, the high school course. He then taught while dedicating his free time to university correspondence work; in six years he had completed degrees in the liberal arts and law. Gerhart notes that Lembede spent some of this time in the Orange Free State, where he became fluent in Afrikaans and took an interest in the fascist writings of future Prime Minister Hendrik Verwoerd, editor of *Die Transvaler*. In 1943 Lembede moved to Johannesburg to apprentice as a clerk with the lawyer and founding ANC member Pixley ka Izaka Seme.

Lembede and the other intellectuals A.P. Mda, Jordan Ngubane, Nelson Mandela, and Walter Sisulu, who would found or play key roles in the new ANC Youth League from 1944, were men of a different generation from the ANC founders. While they may have been educated in the best of liberal institutions, they had witnessed the steady erosion of African rights in the face of Afrikaner legislative initiatives and liberal impotence. Hopes for the generalized extension of Cape liberalism paled for the members of this generation, as the best and brightest among them found decreased scope for the utilization of their intelligence and initiative. Their shrinking hopes were mirrored in the increasingly dire circumstances of Africans generally and this, at a time when the world was speaking of freedom, of liberation from fascist tyranny. The future leaders of the Youth League saw the world rally to the anti-fascist cause, heard world leaders decry the evils of one nation's domination of another. They witnessed Africans leaving for battle in North Africa to fight a war for the liberation of Europeans while shackled, even on these far-flung battlefields by the same restrictions and oppressive measures meted out at home. Nor could this young African intelligentsia help but notice the power of nationalism, at home or abroad. Afrikaner leaders were willing to go to prison in the name of the *volk*, and it was evident to Lembede and others that the Afrikaners' ability to unite their community was a key element in the success of Afrikaner political and economic initiatives. All these factors impinged on the consciousness of Lembede and his colleagues as they struggled to define a philosophical and activist path for the African people.

The creed deriving from the intelligence and experience of Lembede and Mda was known as "Africanism." Africanism shared with its liberal predecessor a belief in the necessity to move from the consciousness of ethnic affiliation to that of a shared and common Africanness. However, where African liberals had looked outward, to salvation through the good will of whites in combination with Africans' western acculturation, Africanists looked to a change in the internal life of the African people, to the creation of a positive self-conscious sense of pride in African history and African being, the beauty of black skin on a black continent. Lembede's ideas sprang from his estimation of the dire psychological straits of the African people in South Africa. They were cowed, had forgotten their history or, worse, repudiated it, and had lost all sense of self-worth. They had been told they were inferior to whites and had come to believe it. Africanism offered an antidote. It returned to the African people their inherent value, as human beings and as Africans in an African land -- Africa was for Africans.[9]

Africanism, as Lembede espoused it, was a potent ideology, but it offered no prescription for organization or action. Yet, by virtue of its recognition that Africans would have to save themselves, it did offer some direction. If liberalism was ultimately a means of co-optation and control, then Africans had to reject it in all its forms. If African leaders had espoused liberal ideas of incrementalism and acculturation, they were to be rejected along with the white liberals themselves. It therefore followed that Africanists would have no use for state institutions or projects promoting "dialogue" with whites or limited African representation. Boycott would become a central theme of the Youth League's program. Africanism told Africans that they were on their own. They would have to save themselves. Gerhart has written that Lembede imputed to Africans a natural activist nationalism.[10] Lembede assumed that Africans tended naturally toward a nationalist, as against a liberal, outlook, and the power of this nationalist consciousness had only to be awakened and tapped. Again, while this principle was not framed as a blueprint for action, it suggested a general direction: to use the demonstrative power of the African masses to confront apartheid authority. Lembede died in 1947, nearly two years before the ANC's Youth League inscribed its Programme of Action. Under the direction of new ANC President Walter Sisulu, the Youth League adopted the Programme of Action, a radical departure from former ANC doctrines:

> The fundamental principles of the programme of action of the African National Congress are inspired by the desire to achieve National freedom. By National freedom we mean freedom from White domination and the attainment of political independence. This implies the rejection of the conception of segregation, apartheid, trusteeship, or White leadership which are all in one way or another motivated by the idea of White domination or domination of the White over the Blacks. Like all other people the African people claim the right of self-determination.[11]

The Programme of Action called for the exercise of extralegal methods, including strikes, stayaways, and boycotts to challenge the state. The Programme of Action would provide a general plan for moving the African people to action, repudiating the go-slow elements within, and challenging encroaching white supremacists and measures without.

Africanism dictated action through independence. Africans would work among themselves, at their own pace, and would not be sidetracked by any other group's agenda. Throughout the 1940s, the Youth League rejected cooperation with either white sympathizers, other black organizations, or the

Communist Party. Youth Leaguers were highly suspicious of any group or ideology that might be considered foreign, not indigenous to African soil. Lembede and others believed that Africans had first to clarify their goals, aspirations, and identity among themselves. Cooperation with liberal whites, Indians, or Coloured South Africans would distract Africans from the task at hand: their own liberation.

The Youth Leaguers reserved some of their strongest venom for the Communist Party. Africanists rejected any cooperation with the Communist Party, given what they perceived as the fundamental contradiction between the party's class analytical approach and the Youth League's nationalism. Of equal importance in the minds of Youth Leaguers was the fact that the Communist Party was a white organization, led by foreigners and spouting an alien doctrine. Africanists rejected cooperation with whites generally, *as* whites. Even those with the best apparent motives held little clout against the majority of their brethren who shared "the spoils of white domination" in South Africa.[12]

While the Programme of Action revolutionized ANC doctrine, one is hard pressed to find in it the rich, almost poetic ideological pronouncements characteristic of Lembede.[13] The Programme of Action was focused on action, and while it made reference to African nationalism as a guiding principle, few of Lembede's central concerns, such as the welding of a strong, self-conscious African nation, could be found there. Indeed, in explaining how a number of the Youth Leaguers of the 1940s became the strongest proponents of multi-racial organization in the 1950s, approving of tight-knit cooperation with whites and communists, their seeming lack of interest in replicating Lembede's most vociferous pronouncements offers an initial clue.[14]

The Communist Party and the National Question

If the most doctrinaire positions held by the communist and Africanist leaderships had prevailed at the end of the 1940s, the coalition building and pan-ethnic cooperation that foreshadowed the ANC-led Congress Alliance of the 1950s might never have occurred. However, a number of leading Africanists found value in tactical alliances and did not allow whatever fears, prejudices, or frustrations they may have harbored to deter them from a broader unity.

Communists, like their Africanist counterparts, recognized self-imposed barriers to participation in the creation of a broad-based nationalist liberation movement. The party would have to square its vision of the working class as the vanguard of the socialist revolution (and the party as the vanguard of the working class) with the alternative vision of a bourgeois national liberation movement led by a petit-bourgeois nationalist movement, the success of which would lead to the emplacement of a democratic capitalist society. The Communist Party would have to resolve the "national question."

Integral to the national question is the question of the class content in national struggle. Is nationalism a form of class struggle or is it an autonomous force? For Marxists, as for Marx, Engels, Lenin, and Luxemburg, national struggle has been understood as a part or form of class struggle. The object of class struggle is the attainment of state power, and national struggle can be seen as one form of this struggle.[15]

Analysts of the national question in South Africa faced a unique situation, as both oppressor and oppressed "nations" existed within a single territorial entity. The Black Republic thesis of the later 1920s had posited a bourgeois democratic revolution of workers and peasants led by the black working class, when most party workers believed the revolution would be socialist in content and driven by the white working class. Throughout the 1930s and much of the 1940s the Communist Party discounted the revolutionary nature or character of the national movement. While the revolution was discussed in non-racial terms, as a revolution of and for the working class, the party's appeals to white workers in the 1930s remained cast in the racial idiom of the time. Communists looked to the revolution to provide for the human needs of all, black and white, but the party's rhetoric sought to accommodate white fears, noting that with the effacement of poverty and backwardness race mixing would be less likely to occur.[16]

Moses Kotane, a leading party member, called, in a 1934 letter to the Johannesburg District Committee, for the Africanization of the party. Kotane criticized party members, and white members specifically, for subordinating their analysis of South Africa's conditions in favor of their interest in European matters. Kotane asked that the party "study the conditions in this country and concretize the demands of the toiling masses from first hand information." He criticized the party as "a group of Europeans who are merely interested in European affairs."[17] The party's involvement in broad front activities from the later 1930s onward might appear to have met these criticisms. However, the party leadership still refused, by and large, to acknowledge that where race and class were synonymous as modes of black oppression, struggle against overtly racial manifestations of oppression such as the pass laws and segregatory legislation was itself revolutionary.

Only in 1950, shortly before the party's dissolution, were there hints that the party's interpretation of the national question was undergoing reevaluation. The party Central Committee's report to the annual CP conference was tabled in January 1950. While the report's prescriptions for action remained very much based upon the initiative of class conscious African workers and peasants, the report stated that the country "is entering a period of bitter *national* conflict" in which the "realities of the class divisions are being obscured." Criticism of the national movements spoke primarily to the weakness of the black bourgeoisie. This weakness, in the party's analysis, left leadership in the hands of the petite bourgeoisie, "teachers, church ministers, professional men," who were incapable of organizing and leading mass movements because they were "handicapped by their dependent economic status." The party document criticized African nationalist formulations calling for "true democracy" and "a just social order" as the freedom of capitalism. The people were misled because the equality they achieved would be bourgeois in content. The Youth League's Programme of Action was attacked as vague and crude. At the time of writing, the strategists in the Central Committee perceived nationalism without an equal component of class consciousness as potentially racist. They stated that nationalism "need not be synonymous with racialism, but it can avoid being so only if it recognizes the class alignments that cut across racial divisions." Evidence of this was the African nationalist leadership's assumption that "*all* Europeans are the enemies of the Non-Europeans, that *no* European can be trusted to fight wholeheartedly all the time for the liberation of the Non-Europeans, and that their liberation can be achieved only through exclusively Non-European organizations."[18]

There seems to be little in the way of a reevaluation of the party's fundamental analysis in any of this. African nationalism is given little credit, philosophically or in the organizations promoting it. There is a sense of hostility and righteousness which might be understood, to the extent that it emanated from white party members, as a response to the threat posed by the Youth League. The Youth League repudiated communism and its white ringleaders. The nationalism of the Youth League thus threatened the participation of whites in the liberation movement, and, by extension, white radical identity as constituted by participation in the revolutionary process. Yet, reading further in the Central Committee's report, there is a certain prescience exhibited in its instructions for the building of the national movements.[19] The Central Committee called for the amalgamation of the national organizations in a revolutionary party encompassing workers, peasants, intellectuals, and the petite bourgeoisie, in alliance with white workers and intellectuals:

> Such a party would be distinguished from the Communist Party in that its objective is national liberation, that is, the abolition of race discrimination, but it would cooperate closely with the Communist Party. In this party the class-conscious workers and peasants of the national group concerned would constitute the main leadership. It would be their task to develop an adequate organizational apparatus, to conduct mass struggles against race discrimination, to combat chauvinism and racialism in the national movement, to develop class consciousness in the people, and to forge unity in action between the oppressed peoples and between them and the European working class.[20]

Undeniably, the emphasis on a class-conscious workers' vanguard remained. Still, there was a clear and dramatic shift in focus from the class to the national struggle, from a primary focus on building worker unity and pursuing workers' interests toward the building of a pan-class, multi-racial movement for the purpose of agitating against national oppression on a mass basis. The party leadership may or may not have foreseen that, in the whirlwind promulgation of apartheid, the Communist Party would be a primary and early victim. The Central Committee believed that the party would be a major influence in the national liberation movement through the participation of "class-conscious workers and peasants," who would "combat" the negative, chauvinistic elements attendant upon nationalist organization. In fact, workers and peasants would not constitute the nationalist vanguard. Communist Party members did, however, play a key role in the national liberation movement of the 1950s, through a number of national organizations, including the ANC. White communists were, perhaps, the most prominent, or visible, representation of communist influence in the national liberation movement, and their presence was a matter of controversy and contention. What role might whites play in a movement for black liberation? Was white liberation an element of African liberation? And how would white communists articulate with a liberation movement whose goals were clearly bourgeois democratic?

These questions, manifestations of the national question, were only resolved from the early 1950s. The answers were embodied in the South African Communist Party's[21] (SACP) conceptualization of "colonialism of a special type" and the two stage theory of revolution which posited the national-democratic revolution as a necessary prelude to socialist victory.[22] These conceptualizations placed the party's analysis firmly in the South African context. South Africa, as a colony of a special type, exhibited the characteristics both of a colony and an imperial metropolitan power, a country in which colonizer and colonized lived and worked next to each other. The two stage theory of revolution in response to South Africa's supposed uniqueness posited a dual liberation process. First, the oppressed majority had to achieve its national liberation on the basis of the broad front. There would be a period of national-democratic rule as a stage toward a working-class revolution and socialism. This enunciation of the role of nationalist agitation came only after the crushing of legal organization between the late 1950s and early 1960s, including the banning of the ANC and the Pan-Africanist Congress. The party moved toward this theoretical position as it regrouped illegally from 1953, in the context of its members' experiences of, and in, the national organizations.[23]

The Ambiguity of White Politicking in the 1940s

The Communist Party enjoyed a spectacular resurgence through the war years. The Central Committee reported a nationwide membership of 1,500 in

1944, a quadrupling of party numbers in four years.[24] With the German attack on the Soviet Union, the party adopted an interventionist stance and spoke unambiguously for war against fascism both abroad and at home. The Soviet Union's valiant stand against the Nazis won the party an unprecedented degree of acceptance in South African society. The party drew on this unusual store of respect to speak for equal rights for black soldiers and to champion basic needs causes for all South Africans.[25]

White party members were particularly visible in these party activities. Party attempts at electoral politics focused on three outlets in particular: municipal elections; African parliamentary representative elections in the Cape; and African location Advisory Board elections. The bulk of party energies and party press attention focused on the arenas (Parliament, City Councils) that whites alone contested. The motives for this trend in party activity were ambiguous: they could be understood as a desire not to make waves that might diminish its acceptance or as evidence of the party's belief that Africans were not ready to be organized for mass action. To some it might have seemed controlling or paternalistic. Yet the fact of the matter was that a small number of party whites, acting as spokespersons for blacks in the institutions of the dominant white culture, gained near heroic status among Africans while speaking out strongly on issues of immediate and crucial concern to blacks. In this way, a small group of whites moved closer to an identification with the most basic elements of black oppression and aspirations.

Electoral Politics: Process

The Communist Party expended a great deal of energy contesting municipal elections in the mid-1940s. They attempted to contest a small number of seats at the parliamentary level as well. Victories were few and far between. Yet there were few signs of regret at the party's expenditure of energy working within the white electoral framework. The great majority of party candidates stood either in Johannesburg or Cape Town. The Johannesburg city electorate was white only. In Cape Town Coloureds who met certain property qualifications still held the vote, and Africans maintained the right to a representative in Parliament and on the Cape Provincial Council (CPC).

The primary contradiction the party faced in its electioneering efforts among white constituencies was one of definition. Party candidates would be appealing largely or exclusively to white voters. But in whose name? White workers? All workers? Africans and other blacks? All South Africans? The signals sent by the party were, taken as a whole, mixed. However, if a thread runs through the party's campaign rhetoric, it is the attempt to raise issues critical to African standing and African welfare. Party leaders were highly conscious of the general antagonism toward its policy of non-racialism and of white workers' negative reactions to it. Yet in 1943, with election activity underway, the editors of the party's major circulation paper[26] used the paper's column, "The Communist Party Answers Your Questions," to answer directly those who opposed the party's "race-mixing" policy, first raising the objections themselves:

> Many Europeans say this. They object to non-Europeans living in the same streets as whites. They don't like to sit next to them in buses or trams. They object to the Communist Party because it makes no distinctions of colour, race or creed in its membership. They are horrified because White and non-White people are allowed to mix together at Communist Party dances and socials....We Communists know quite well that this is not a popular view to hold. We know that it is a hard tack to persuade Europeans that this is so. But because we are firmly convinced that this is the truth and that it is morally correct, we fight for the principle of no colour-bar.[27]

The writer does not turn to justifications based on class solidarity or describe racial harmony as a concomitant of a socialist state. Rather, the writer utilizes an analogy common to the 1940s and 1950s in leftist circles, that of a sinking ship, to suggest that neither white nor black could be "saved" from the economic and social ills of the society without recourse to each other. Vague moral formulations such as those above, or the plaint at the end of the column that "South Africans must realize that there is no other way out of their troubles" suggest a certain irresolution in the party's embrace of non-racialism. These formulations exhibit an understanding of the most subjective evils of *baaskaap*, (an Afrikaans term, "boss-ship" connoted the use of legislative and coercive force to maintain white rule) the objectification and dehumanization of Africans. But the analysis of these evils is divorced from the party's usual materialist framework. Segregation was, simply, wrong.

Communist candidates included with varying degrees of explicitness discussion of issues and needs central to African welfare. These discussions and exhortations could be as general as a call for "free clinics in all areas" or "clean sanitary conditions for all sections of the population." Such coded language spoke to the particular hardships faced by African squatters and shanty dwellers on Johannesburg's periphery. Rhona O'Meara, a party candidate for the Johannesburg City Council in 1944, suggested a similar agenda in describing her platform as that of "considering the interests of the people before profits, and by tackling evils by removing the cause. Only by fighting for all sections of the people, irrespective of race or colour, can Johannesburg be made into a model city."[28] O'Meara included more specific references to this general theme, noting that Orlando Township's 60,000 residents enjoyed the services of but a single part-time doctor while the profits from state-run township beer halls funded municipal projects in white areas.[29]

Franz Boshoff, a distinguished advocate, was a descendant of a Free State President. Boshoff had grown up in the liberal Standerton district of

Johannesburg and was an outspoken student at the University of Pretoria. He joined the Communist Party in 1938 and stood for a Johannesburg City Council seat in 1945. Boshoff targeted his Afrikaner brethren during the campaign but spoke as forcefully for the rights of Africans as he did for the plight of poor whites:

> As communists we demand better health services for the non-Europeans, not simply to protect the health of the European community, but because we believe in the fundamental right of all people to enjoy healthy living conditions....There is a vital necessity for the poorer sections of our population, particularly the non-Europeans, to have someone to fight for better conditions for them through the City Council.[30]

Boshoff thought it essential that communists be elected and fill that role. Yet, one must ask: What did white party members seeking seats in segregated municipal institutions expect to accomplish? They probably had in mind a handful of goals. A 1944 party post-election report suggested that municipal elections offered "the best possible propaganda for our party and our ideals."[31] Ward elections provided excellent opportunities for getting out the party's word street by street. And an important component of that word was pro-Soviet propaganda. At a time when the Soviets were an official South African ally, the idealism of South African CPSA members could take wing in voice and print. Hilda Watts, in a successful campaign for a 1944 Johannesburg City Council seat, wrote in her election manifesto:

> We all want a future of security, expanding democracy, and freedom from want and fear. Only in the Soviet Union, under the Communist Party of that great country have the people really attained these human ideals. Guided by the same lofty principles, the Communist Party can build a healthy, happy Johannesburg on the path to a future of socialism and true democracy in South Africa.[32]

White party candidates used elections to publicize the party's existence, ideals, and platform; to propagandize for the Soviet Union, which many believed had achieved a higher form of political, social and economic organization; and, of equal if not greater importance, to speak for Africans and other disenfranchised blacks where they had no voice of their own. Perhaps the war-time atmosphere had kindled whites' consciousness of the evils of racial oppression. Whatever the cause, there was a sense within the party in the mid-1940s that the white public could be swayed toward providing a more favorable dispensation for Africans. During the 1944 bus boycott in Johannesburg, a Johannesburg District Committee report stated that there was "a large and growing body of European opinion conscious of the oppression of the non-European people, and sympathetic to their demands and aspirations. It is the special duty of every

European Communist devotedly and energetically to spread and deepen this consciousness and sympathy."[33]

Communist Party Council candidates, particularly those in Johannesburg, were faced with a contradiction. They were attempting to speak about African needs and disabilities to an electoral base that was exclusively white. At its most vociferous, the candidates' rhetoric counted white voters as a means to win the right to use city funds to aid blacks. It was a platform that won the party the respect of many Africans. However, it won the party few elections in 1945 or at any other time. The white South African public not only distrusted communists generally, but did not identify with the party's goals. The white working class might appreciate the party's general formulations on the need for more housing, better health care, or enhanced social amenities. However, when party candidates demanded "the extension to African and other non-European peoples of all citizenship rights, including the democratic right to participate fully and equally in the country's law-making bodies," they immediately lost the majority of their potential constituency.[34]

The Communist Party continued to put up candidates for election in the years between the end of the war and the 1948 general election. The party faced new hardships as ward redistricting robbed it of concentrated progressive voting blocs. In April of 1946 CPSA candidate Harry Snitcher lost Cape Town's Castle ward contest when part of the ward's Coloured voting population was lopped off, leading to a United Party victory. Michael Harmel, a leading party member, received 690 votes to his opponent's 6,000 plus votes in Johannesburg's tenth ward as the result of similar redistricting. The party decided against contesting a 1947 Provincial Council election because of government attacks on the party, including arrests and confiscations of party documents.[35] Still, in the heart of the 1940s, the party did enjoy a few victories that placed party members on the city councils of South Africa's largest cities and in parliamentary seats as African representatives.

Communist Party Candidates: Taking Office

What was the effect of white Communist Party members holding public office? Communists in city councils had few allies if any and cooperation with Labour blocs tended to be informal and tenuous. Nonetheless, the few communists who did sit on city councils and other public bodies had an impact on the proceedings of those bodies. Hilda Watts, who served on the Johannesburg City Council during the war years, and Sam Kahn, who sat both on the Cape Town City Council and in Parliament as a Cape African representative, were among the party's most successful contenders.

Hilda Watts was born in London but had relatives, including an aunt, in Johannesburg, where Watts went to live for some time in the early 1930s. Returning to England, she took an interest in politics and joined the Young

Communist League. Watts returned to South Africa in 1937, where she participated in Communist Party work. She returned to England once again before settling in South Africa. Watts and other young party members had also participated in the Labour League of Youth (LLY), the Labour Party's youth wing, in the hope of influencing that body, until a decision was taken within the party that members with dual CP/LLY membership should declare for one organization or the other. Watts declared her membership in the Communist Party.[36]

During her first campaign in February 1944 Watts' manifesto addressed basic needs, such as the provision of homes, rent control, play areas for children, and health care for all Johannesburg citizens. Watts noted the party's full support for South Africa's part in the war effort and demanded that homes be provided for returning soldiers. Watts lost this election. Running for the same tenth ward seat just eight months later, she won the election, becoming the Communist Party's first elected representative in the Transvaal. During the campaign she mentioned the surprise she had encountered among the public at the reformist rather than revolutionary nature of the party's electoral program. "Our Party," she said, "has always been concerned with the local small problems of the people....We are concerned with local problems in connection with the larger issues which take place in this and other countries."[37] Hilda Watts was an energetic and outspoken party representative. If some potential constituents were surprised that a party candidate evinced knowledge of, and concern for, local bread and butter issues, they might have been equally surprised at Watts' demeanor on the council floor. Watts took every opportunity to challenge the conventional racist wisdom prevailing there. At the time of her first victory Watts declared that she would represent all the people of Johannesburg, "including those who have not yet the privilege of choosing their own councilors."[38] Thus, Watts announced a dual agenda, representing the interests of both her white working class constituents, and the underrepresented Africans, Indians, and Coloureds of Johannesburg.

Watts' activism on behalf of her shadow African constituents was manifest almost immediately. In December 1944, six Johannesburg City Council members presented a petition to the council demanding removal of all Africans from Newlands, Martindale, and Sophiatown. The petition's seconder, M.C. Schoeman, dismissed the western townships as "a breeding place of drunkenness and misdeeds."[39] Watts, in reply, described the conditions of the African population there, the poverty, illness, and atmosphere of mounting grievance and resentment. She stated that the African leadership in the western townships did not approve of the removal scheme and did not wish to move. Watts' defense was powerful, but she was a lone voice in the city council, and so merely a thorn in the side of the council majority. Still, Watts did not shrink from confrontation. When a resolution was introduced (in the city's General Purposes Committee) for the purchase of advertising space in an African newspaper to publicize the council's efforts on behalf of Africans, Watts suggested that the best advertisement would be the provision of sewerage and electric light.[40]

In May 1946, the Johannesburg City Council named Johannesburg a "proclaimed" area under the Urban Areas Act, giving the government the right to set highly restrictive conditions for African entry into the city. Watts, in characteristic style, spoke against the legislation as "one more act of injustice" sure to increase African bitterness. Watts was one of three dissenting voices that day, joined by a Labour Party representative and an independent.[41] Even in the relatively legalist atmosphere of the mid-1940s, the city council had means of repudiating enemies from within. In August 1946, Watts was notified that the boundaries of her ward were to be revised, cutting out flat dwellers and small homeowners and adding areas where large single-family properties were located.[42]

Watts continued to fight for the disenfranchised in her last year as a councilor. When Africans in Moroka and Jabavu struggled with the municipality over skyrocketing rents, Watts put a motion calling upon the council to open negotiations immediately with African leaders in those townships. No seconder stepped forward. Toward the end of her council tenure, Watts attended a meeting called by the African Parliamentary representative Hyman Basner to air African grievances in front of the manager of Johannesburg's Non-European Affairs Department, the chairman of the council's Non-European Affairs committee, and Margaret Ballinger, a liberal spokeswoman and the Eastern Cape "Native" representative. In her statement before the assembled, Watts spoke directly to Johannesburg's African population. Her remarks are interesting both for suggesting Watts' own estimation of her impact on the council, and for their prescriptive sense:

> As long as you have no political voice you will be hidden away in Morokas. If Moroka Township consisted of Europeans I think they would do what you are doing. I am not here to tell you what to do. It is not true that white agitators tell the African people what to do. The agitators are the shacks of Moroka Townships. The babies who die so soon after being born.[43]

There is irony in Watts' statement: the recognition that Africans had no political voice coupled with an admission that Watts' presence could never constitute that voice. She denied the stock anti-communist charges that white communists controlled African political initiatives even while suggesting that Africans would never enjoy rights and comforts if they did not organize themselves and agitate. The Africans of Moroka and other Johannesburg townships would have been hard pressed to find among any racial or ethnic group another spokesperson as forthright and driven as Hilda Watts. Yet the fact remained that she was not an elected African representative, and as a white communist woman, she was an anomaly in Johannesburg city government in the 1940s.

If the Cape Town branch of the Communist Party had a counterpart to Hilda Watts, Sam Kahn was that person; he was a lawyer and a leading party member. Kahn's first Cape Town electoral victory, in the East Central ward, led Bill Andrews, a founding International Socialist League (ISL) and CPSA member, to exclaim that "Twenty-five years of work against bitter and adverse propaganda had been rewarded."[44] In 1946 Kahn was again voted on to the council, this time in the Castle constituency. A dynamic and energetic man, Kahn had taken a B.A. and a law degree from the University of Cape Town while serving as president of the Debating and Law societies and chair of the Students' Jewish Association. Kahn had joined the Party in 1932 and had been a member of the National Liberation League (NLL) and Anti-Fascist League (AFL). Like Watts, Kahn had attempted to swim upstream against a solidly hostile council majority. When the council prepared testimony for the Native Laws Commission, Kahn demanded amendments that would give Africans the "democratic right of freedom of movement [and] freedom of residence" and that would see "the abandonment of locations."[45] Kahn's -- and the party's -- most significant electoral victory was his seating in the House of Assembly as African representative for the Western Cape in 1948. Kahn stated during the campaign that "a vote for a Communist was a vote for freedom. That is what the Africans think today."[46] The Africanists notwithstanding, the election results seemed to bear Kahn out. Kahn's closest opponent, the United Party candidate, received 754 votes to Kahn's 3,780.[47] Lawrence Magweba, writing from the Transkei some months after the election, found that Kahn's victory had "inspired Africans with a desire for Communism," and he suggested that this time should be called "the period of Sam Kahn."[48]

As the African Parliamentary representative for the Western Cape, Kahn used his position both to push the local interests of his constituents and to speak on matters of national significance. Kahn led a deputation from Langa to talk with Prime Minister Verwoerd about rent increases. When Parliament debated the formal annexation of South West Africa, Kahn called it a "naked measure of annexation or incorporation."[49]

As the lone member of the Communist Party in Parliament, Kahn had the dubious honor of speaking against the Suppression of Communism legislation. Perhaps it was a stronger sense of legalism still extant in 1950, or a belief that Afrikaner nationalism might still be defeated at the ballot box, that led Kahn to speak vehemently against the Bill in January 1950.

The manner in which Kahn chose, publicly, to interpret the thrust of the Suppression of Communism legislation suggested that changes had taken place in party thinking throughout the 1940s. When Kahn rose to speak against the Bill, he spoke of the uniqueness of the party, but he did not define this uniqueness in terms of the party's proprietary interest in Marxism-Leninism or in its uniquely vigorous and outspoken defense of the interests of workers and blacks:

> We know what the honorable Minister has against the Communist Party. It is a unique political party in South Africa, because it is the only political party in which Europeans and non-Europeans may be members on an equal footing. There is no other political party in South Africa which accepts non-Europeans as members....The Communist Party has an incorruptible racial policy and it is because the Communist Party is a mirror of the South African society of the future, where all our different racial groups and elements will play an equal part in the Government and the administration of the country, where we would have eliminated the conception of two nations, the one the ruling nation and the other the oppressed nation.[50]

The vision in the latter half of Kahn's statement is without socialist allusion: Its revolutionary nature derives from a vision of national liberation and its aftermath. In that spirit, Kahn's statement represented a step toward the party's resolution of the national question. The resolution Kahn suggested was not that of worker control. It was the resolution of two nationalisms based upon the breakdown of racial barriers. Although the Communist Party's analysis generally defined racial oppression as an adjunct of class forces, Kahn accepted the reality of race-based oppression and the primacy of fighting that oppression on national terms. The conscious organization of black and white on a basis of equality, a contravention of the basic underpinnings of white South African society, was now defined by the party as a revolutionary act.

The Communist Party's involvement in electoral activity was short-lived but was not without effect. The party, through its candidates and elected officials, charted the inequities of South African society and attempted to wrest some justice from that society as apartheid was named and its cornerstones put in place. The party propagandized for communism and for the Soviet state as its laboratory and exemplar. Yet the party won adherents by virtue of the outspokenness of its candidates on issues of immediate interest, not for its Soviet propaganda. Party candidates, white party candidates, spoke for workers, the poor, and the disenfranchised majority.

This role of "speaking for" was problematic, particularly given that whites were in visible and vocal roles. Communists were vulnerable to charges of piracy, of attempting to take control of, and direct, African and other nationally based organizations. And the visibility of whites, of Wattses and Kahns and Harmels, provided fuel for such charges, from Afrikaners, Africanists, liberals and others. Notwithstanding the validity or fallacy of these charges, the party was greatly changed from its 1940s incarnation. Speaking in Parliament at the time of his expulsion from the Cape Western seat in 1952, Sam Kahn opined that:

> [W]hat has brought me into conflict with this government has not been my belief in socialism or my belief in a republic, but it has been my advocacy of complete equal rights for black and white in this country. That is what I am being tried for, and they wish to make that the modern blasphemy, the twentieth century heinous crime in politics, namely, the advocacy of equal rights for black and white in this country.[51]

As proponents of socialism, members of the Communist Party were not difficult to marginalize. There was little interest in the party's message among the one constituency -- the white working class -- that had some potential political power with which to respond to it. The party had miscalculated in its tendency to view the white working class as a force that would ultimately be motivated by unified class interests above all others. However, by the early 1950s, as proponents of equal rights, universal democracy, and integration, communists were dangerous heretics who might well find a place alongside or within the national organizations and so pose a serious threat to the state.

Toward an Integrating Ideology: Multi-Racial Cooperation in the 1940s

There were two outstanding trends or tendencies in the politics of liberation during the 1940s: one was the rise of a militant and self-conscious African nationalism; the other, the solidification of an ethic of cooperation among organizations across racial, ethnic, and ideological lines. The latter trend was crucial for whites seeking to play a role in opposing white supremacy, as it provided the structural underpinnings of the Congress Alliance and the ideological prerequisites for the Freedom Charter. With the Communist Party on the verge of becoming illegal, the organization of opposition along national lines may have seemed a decidedly retrograde trend. Yet it would afford whites, and specifically white communists, the opportunity to continue their participation in, and influencing of, the politics of liberation through a white body complementing the African, Indian and Coloured Congress Alliance affiliates, the Congress of Democrats (COD).

Karis and Carter have noted the "extraordinary complexity" of the efforts to build "non-European" unity or "radical popular fronts of all races, including whites" from the late 1930s through the end of the 1940s.[52] How does one account for the working unity among diverse organizations in the 1940s and beyond? Certainly, leaders of the relevant political organizations in the 1940s had models to look to. The South African nation was born out of a "national convention" of Afrikaner and British representatives. That initial conclave had been held up since Union as the visible fruit of unity among former antagonists, and there were lessons in it for those who sought to redefine the nation. The war period in the 1940s, during which the efficacy of the United Front was trumpeted and realized in action, offered further evidence of the efficacy of combination. The National Party (NP) victory and the influence of the Youth Leaguers on protest politics brought about a certain consensus of method in organizations ranging from the ANC to the CP to the Indian Congresses.[53] The racial violence of the later 1940s, such as the 1949 Durban riots, proved a further spur to communication across racial and ethnic lines. These diverse forces and influences moved the equally diverse organizations battling white supremacist initiatives toward closer cooperation throughout the decade.

Early attempts at alliance-building engendered substantial controversy, and the question of whether whites should be included in pan-racial and pan-ideological alliances was itself a highly controversial element. James LaGuma favored barring whites from participation in the National Liberation League and was barred himself. The Communist Party, which had been calling for organization on the basis of a United Front for a number of years by the mid-1930s, appealed for unity and class solidarity and against "inverted racialism."[54] The Non-European Unity conference in April 1939 represented the apex of cooperation in the 1930s, with organizations as diverse as the Communist Party, the Indian Congress and Trotskyites in attendance. The Simonses, writing of the Non-European United Front (NEUF), said that the "seed of a grand non-racial alliance had been planted, but seventeen years were to pass before it bore fruit," referring to the solidification of the Congress Alliance.[55] Shortly after the NEUF's demise, the Non-European Unity Movement (NEUM) was formed from the coalescing of All-African Convention (AAC) supporters and Anti-Coloured Affairs Department (Anti-CAD) members. The Non-European Unity Movement attempted without success to ally with the Indian Congress and the ANC, but it was antagonistic to the presence of whites and communists in liberation activities.

The major campaigns of the early 1940s, the Anti-Pass Campaign from late 1943 and the Campaign for Right and Justice initiated a month later, continued to unite diverse voices. One Anti-Pass meeting in 1944 found Donald Molteno, a prominent liberal, H.A. Naidoo, an Indian Congress leader, Hyman Basner, an African representative and former Communist Party member, and Moses Kotane, a leading CPSA member, sharing the platform. Kotane's call for "all progressive forces to join the struggle for the abolition of these repressive measures" summed up neatly the prevailing will to create a stronger and more persuasive force through unity. The Campaign for Right and Justice was chaired by the Reverend Palmer, Dean of Johannesburg. Among the Provisional Committee members were African, Indian, and white communists, African and white union organizers, a well-known white liberal, and a number of religious figures.[56] The Campaign drafted a program calling for "Freedom from fear" and affirming the right of all South Africans to the rights and principles expressed in the Atlantic Charter. The *Guardian*, in incendiary language typical of the party, called for the mobilization of the oppressed in struggle, stating that "indignant speeches are not enough to break these chains."[57] Through the end of the war, there were quite a few indignant speeches as products of these unity campaigns but little action. Still, for those who believed, or were coming to the belief, that only a coalition of all progressive forces could challenge the status quo, these campaigns were to prove fruitful in the longer term.

The next major campaigns arose only in the last years of the decade. Yet there were abundant calls for concerted action, and proponents of unity pleaded that citizens transcend their old hatreds and fears. Writing in 1945, an African correspondent deplored the unwillingness of Africans to work with Coloureds, Indians with Africans: "The truth is that if you want to win your case please hideaway [sic] the hatchet of race hatred and form a combined front to fight together against whatever unacceptable policy or act is thrown in your midst."[58] There were whites who were also learning lessons about unity, if in different venues and contexts. In 1946, when protest against the "Ghetto Bill" had begun, a white man was arrested with four Africans in an illegal protest in favor of Indian rights. That man, A.H. Mayer, manager of a Johannesburg cinema, traveled to Durban to protest, he said, because of the government's attitude toward the men with whom he had fought "for liberty and democracy": "During the time I spent on the battlefield I mingled with many Indian and Bantu soldiers. We fought for what we believed to be just at the time....[On returning] we found the colour bar, racialism and oppression."[59]

It was all well and good that individuals came to these points of view. How does one account, however, for the fact that the ANC, coming into its own as the voice of Africans during this period, did not turn to the rhetoric of racial division, particularly given the rise of Africanism during this period? The realities of economic interdependence, the ANC's historical commitment to racial cooperation, the abiding influence of Christianity upon African communities and continued contact with white liberals mitigated the possibility of a turn to the rhetoric of division.[60]

The experience of the war, with its democratic rhetoric and universalist themes, also contributed to the development of the ANC's outlook and its relationship to other organizations at this time. The ANC gave its support to the Indians' passive resistance against the Asiatic Land Tenure Bill, that threatened to dispossess the Indian community. At a large anti-pass rally in Johannesburg's Market Square, Dr. Xuma declared that "the battle of the Indians is the battle of the Africans."[61] The African National Congress-South African Indian Congress (SAIC) alliance was formalized in the so-called Doctors' Pact in March of 1947. Dr. Naicker described the Pact as a "historic development," and in the same speech noted that the policy of the Indian Congress was "in no way anti-European....we stand for friendship with the European community."[62]

Shortly thereafter, a meeting was held between representatives of the ANC, SAIC, and the African People's Organization (APO, formed after the South African War, with a mainly Coloured membership, in response to perceived betrayal by the British). M.D. Naidoo, later speaking of the earlier unity meeting, was at pains to point out that while blacks were making common cause as a result of common deprivation, they "would welcome the friendship and assistance of democratic Europeans."[63] But in what guise and under what conditions? Whites sought involvement in the rising tide of nationalist activity, and others encouraged that involvement. Yet, if there were African organizations, and Indian organizations, and Coloured organizations, there were no white organizations, even if some critics would have described the Communist Party as exactly that. Whites came as communists or liberals or religious figures but shared no "common deprivation," at least not in the physical sense. Their presence thus appeared somewhat anomalous.

To create a truly inclusive liberation movement, the primary participants had to move beyond the notion of deprivation as the sole organizing principle and badge of inclusion. The re-creation of the South African nation based upon universal principles of human rights and respect, and achieved through democratic means, would allow whites a positive role. The idea of the national convention, discussed earlier and an important thread in South Africa's history, facilitated the vision of a more inclusionary, democratic South Africa to come.

The National Convention Idea and the People's Assembly Campaign

In late August 1947 Yusuf Dadoo announced at the Transvaal Indian Congress' (TIC) conference that a joint committee of the ANC, Natal and Transvaal Indian Congresses, and the APO were calling a national conference of all progressive organizations at which a "charter for democracy" would be drawn up. Dadoo evinced the hope that:

> It may be a prelude to a national convention truly representative of the South African people irrespective of race or colour. This national convention will, I hope, become a people's parliament to bring about radical changes in the body politic of South Africa in order to remove all causes which are leading this country to ruin, and to win democracy for all.[64]

As the war was ending, the Communist Party had led the call for a national or people's convention. At its 1945 conference the party's Central Committee was instructed to initiate a "People's Convention of the CP, LP [Labour Party], Trade Unions, non-European liberatory organizations and of all progressive bodies and groups, with a view to securing united political action."[65] The party's goals for such a conclave were relatively limited: nationalization of the mines, the end of employment color bars, state control of prices and production, and generalized provision of social services. Hyman Basner, Transvaal and Free State African representative in the Senate, called for the convening of a national convention from the floor of that body. Basner found no sympathy in the Senate itself, but black reaction was positive. Xuma of the ANC, Mofutsanyana of the CPSA, and Cachalia of SAIC welcomed the call, and even Anton Lembede, the brilliant first president of the ANC Youth League (ANCYL) and an ardent anti-communist, described the convention idea as a potential "good step," providing no organization arose from the convention itself that might attempt to supersede those already in operation.[66] The idea of a truly representative national convention appeared infectious by the late 1940s.

The fruit of this convention momentum was the last major campaign of the 1940s, the People's Assembly for Votes for All. The People's Assembly was born in controversy. Some ANC members considered extra-organizational work superfluous, while others, including some Youth Leaguers, would participate only if attendance was limited to ANC, TIC, and APO members.[67] Still, when the conference of the Assembly took place in late May 1948 the 300 delegates were representative of all South Africans across a broad ideological spectrum. A number of the Campaign's aspects suggest its importance as a precursor of the Congress Alliance in the 1950s. The People's Assembly was to be a gathering based upon delegate selection and was to be representative of all the disenfranchised. The Assembly published a pre-conference "Manifesto" that called for a number of delegates sufficient to represent more South Africans than those "voting in the general elections."[68] Delegate selection was similar to that of the Congress of the People, as volunteers went into the factories, hostels, and townships and spoke to sporting, religious, political and other organizations, urging them to send representatives to the Assembly. Teams of volunteers were also sent into the countryside in the Transvaal and Free State.

The Assembly was convened in June 1948. The 322 delegates articulated a "People's Charter" demanding the right "to stand for, vote for, and be elected to all the representative bodies which rule over our people."[69] The People's Charter and the Freedom Charter, adopted seven years later, were similar only in the most general of senses. The most evident similarity was each document's challenge to the status quo, redefining the boundaries of full South African citizenship. The Assembly also called upon the executives of the ANC, SAIC, and the APO to plan "a National Assembly of the South African peoples."[70] Shortly thereafter, Prime Minister Malan unveiled the Nationalists' initial apartheid initiatives, including the end of African representation, the striking of Coloureds from the Cape common roll, and the imposition of apartheid on the Cape rail system. At a Communist Party-sponsored Johannesburg rally in September 1948 the assembled adopted a resolution calling on ANC, APO, SAIC, and Trades and Labour Council (T&LC) leaders to convene, with the party, a "national convention of democrats of all races" to continue the process initiated by the People's Assembly.[71] Subsequently, the Natal Indian Congress (NIC) called for the formation of a "united democratic front of all anti-fascist forces."[72] In the last years of the decade the demands for the formation of a unified anti-apartheid force were legion. The chair of the Port Elizabeth ANC and CPSA member Raymond Mhlaba, speaking in the context of the loss of the Coloured vote, the Durban riots, and the Votes for All Campaign, described this time: "[A]s an opportunity for us to join forces -- the Africans, Coloureds, Indians and sympathetic Europeans -- under the slogan, Votes for All....We suffer under the same disabilities. Our fight is not directed against the European population, but against the Government."[73]

Explicit in these calls for unity was the message that non-Africans had nothing to fear from African nationalism. James Moroka, the ANC's new president in 1950, declared that the ANC fought not only for African freedom but for Indian and Coloured freedom and that the ANC was willing to "join hands even with those Europeans who are prepared to fight with us -- and there are many of them."[74] The fact remained that, at the time of Moroka's statement, the whites who stood most prominently with the national organizations in calling for full democracy were, largely, communists or ex-communists like Hyman Basner. The Communist Party was shaped throughout the 1940s by a varied array of forces and influences. The visible increase in nationalist sentiment and political and labor-related agitation among Africans was critical in moving a working class party to appreciate the possibilities of broad-based coalition building. Local struggles such as bus boycotts and rent strikes and the radicalization of the Indian community moved the party toward acceptance of a national liberation framework. Yet the party's analysis of the South African political scene remained focused on the working class, de-emphasizing race as an adjunct to capitalism in the South African equation. The party's mission, in the context of the national organizations, was to show that "the colour bar is primarily a technique of exploitation for private profit, by emphasizing the unity of interests that exists between the workers of all races, and by ensuring the dominant role of the class conscious workers in the national organizations."[75]

The imposition and elaboration of racially based laws and structures may not have led white communists to a wholly new appreciation of racial ideology as a serious and independent problem. However, the progress of events in the 1940s did lead white party members to acknowledge the need for an immediate response to the Nationalist onslaught on a national rather than a narrowly class basis. The party did not abandon its agenda: Rather, it changed priorities, making national liberation its primary project and hoping to influence the national organizations to adopt its analytical framework through work within those organizations. This, in turn, led party members to work with the national organizations closely in pursuit of liberation.

Notes

1. David Everatt, "Alliance Politics of a Special Type: The Roots of the ANC/SACP Alliance, 1950-1954," *Journal of Southern African Studies*, Vol. 18, No. 1 (March 1991), 25. The evolution of the South Africa-specific theory of "colonialism of a special type" at the hands of party theoreticians like Michael Harmel and Jack Simons served to provide a resolution of sorts to the national question, bridging the national and class struggles.

2. *Guardian*, 2 April 1947.

3. Ibid.

4. *Guardian*, 1 May 1947.

5. Gail Gerhart and Thomas Karis, *From Protest to Challenge: A Documentary History of African Politics in South Africa, 1882-1964*, Vol. 4, *Political Profiles, 1882-1964* (Stanford: Hoover Institution Press, 1964), index.

6. Ray Alexander, "Our Party Must be Made a Leading Force," *Freedom*, No. 10, July 1942.

7. Moses Kotane, *Freedom*, September 1942.

8. Gail Gerhart, *Black Power in South Africa: The Evolution of an Ideology* (Berkeley: University of California Press, 1978), 51-54.

9. Ibid., 67.

10. Ibid., 78.

11. African National Congress, "Programme of Action", adopted at the ANC Annual Conference, 17 December 1949, in Thomas Karis, *From Protest to Challenge: A Documentary History of African Politics in South Africa, 1882-1964*, Vol. 2, *Hope and Challenge, 1935-1952* (Stanford: Hoover Institution Press, 1964), 337-339.

12. Ibid., 72.

13. Ibid., 84.

14. Ibid.

15. This general discussion of the content of the national question is indebted to James M. Blaut, *The National Question: Decolonising the Theory of Nationalism* (London: Zed Books, 1987), 23.

16. *Communism and the Native Question* (pamphlet), Communist Party of South Africa, South African Institute of Race Relations (SAIRR) Political Documents, 1929-1972, Institute of Commonwealth Studies, Microfilm 860.

17. Letter, Moses M. Kotane, 23 February 1934, Doc. 55, *South African Communists Speak, 1915-1980* (London: Inkululeko, 1981), 120-122.

18. "Central Committee Report to the national conference of the Communist Party, Johannesburg, January 1950," Doc. 91, *South African Communists Speak*, 200-211.

19. Everatt, "Alliance Politics of a Special Type," 20. By the late 1940s Everatt notes that the Communist Party in the Cape was warning of the dangers of African nationalism and called upon the party to hold fast to class struggle. In Johannesburg, where party leaders were fully exposed to African nationalism and its progenitors, there was less antagonism to African nationalism, and it was there that the theory of colonialism of a special type evolved.

20. Ibid.

21. With the revival of the Communist Party as an underground organization in 1953 the name was changed from Communist Party of South Africa to South African

Communist Party to suggest the newly acknowledged change in emphasis and direction.

22. "The Road to South African Freedom," program of the South African Communist Party adopted at the fifth national conference of the party held inside the country in 1962, Doc. 115, *South African Communists Speak*, 284-319.

23. Everatt, "Alliance Politics of a Special Type." Everatt, writing of the decision to disband the CPSA in 1950, suggests that the decision, far from being the product of grassroots consensus, was hastily made, leading to confusion and disillusionment among many party members. Many of those members believed the party's dissolution was a ruse, and the party's rebirth would follow shortly thereafter. In actuality, much of the Party leadership was not interested in operating illegally. Everatt suggests that the South African Communist Party should not be understood as a simple reconstitution of the old CPSA. Rather, it was a response, among Transvaal members, to the party's absence and a fear of ensuing factionalism, and signaled a major change in party direction, with a relocation of the party's center to Johannesburg and a Johannesburg leadership which took a much greater interest in the emerging Congress movement than had its Cape Town counterparts.

24. *Guardian*, 20 January 1944.

25. Lodge mentions this and it is discussed below. See Tom Lodge, *Black Politics in South Africa Since 1945* (London: Longman, 1983), 28-29.

26. The Party's paper, the *Guardian*, had an average mid-1943 circulation of 42,500 and was edited at the time by Betty Radford (Sacks). Radford came to South Africa in 1931 and held a number of jobs, including typist, hotel worker, and as a publisher's agent on the *Cape Times*. In 1935 she traveled to the Soviet Union for three and a half months, after which she said that the visit "determined me to work for socialism with all the energy I have." She joined the Communist Party in 1941.

27. "The Communist Party Answers Your Questions," *Guardian*, 20 May 1943.

28. *Guardian*, 19 October 1944.

29. *Guardian*, 14 September 1944.

30. *Guardian*, 25 October 1945.

31. "Report of the District Committee to the Conference," Communist Party of South Africa, Johannesburg District Committee, April 1945, SAIRR Political Documents, Microfilm 860, Institute of Commonwealth Studies.

32. *Guardian*, 17 February 1944.

33. "Alexandra Bus Dispute," Johannesburg District Committee Report, Communist Party of South Africa, April 1945, SAIRR Political Documents, Microfilm 860, Institute of Commonwealth Studies.

34. *Guardian*, 8 January 1948.

35. *Guardian*, 9 January 1947.

36. Interview, Hilda Bernstein, Dorstone, 25 January 1987.

37. *Guardian*, 7 September 1944.

38. *Guardian*, 2 November 1944.

39. *Guardian*, 14 December 1944.

40. *Guardian*, 2 August 1945.

41. *Guardian*, 6 June 1946. Ben Weinbren, the Labour councilor, was a prominent unionist who had helped to build up the Federation of Non-European Trade Unions and had attempted to fight fascist tendencies in predominantly Afrikaner union organizing through worker education. A member of the Communist Party, he was expelled with

S.P. Bunting and other Communist Party members in September of 1931, and subsequently joined the South African Labour Party. The Independent councilor mentioned was Joyce Waring.

42. *Guardian*, 15 August 1946. Watts lost half of Berea, and Parktown. Westcliffe and Killarney were added.

43. *Guardian*, 31 July 1947.

44. *Guardian*, 9 September 1943.

45. *Guardian*, 13 March 1947.

46. *Guardian*, 11 November 1948.

47. *Guardian*, 25 November 1948.

48. *Guardian*, 19 May 1949.

49. *Guardian*, 3 March 1949.

50. *Guardian*, 29 June 1950.

51. *Clarion*, 29 May 1952.

52. Karis, Vol. 2, *Hope and Challenge*, 107. This introductory essay provides a useful chronological sketch of the twists and turns toward unity among these organizations, including the African National Congress, and the South African Indian Congress in the late 1930s and 1940s.

53. Peter Walshe, *The Rise of African Nationalism in South Africa: The African National Congress, 1912-1952* (Berkeley: University of California Press, 1971), 357-359.

54. Simons and Simons, *Class and Colour*, 503-505.

55. Ibid., 504.

56. *Guardian*, 14 October 1943.

57. *Guardian*, 16 December 1943.

58. *Guardian*, 4 January 1945. Mayer's comment is interesting not only for his willingness to make it in print, but for the fact also that he writes as if one would not have determined these to be attributes of pre-World War II South Africa.

59. *Guardian*, 17 October 1946.

60. Walshe, *Rise of African Nationalism*, 368-371.

61. *Guardian*, 8 August 1946.

62. *Guardian*, 5 June 1947.

63. *Guardian*, 2 April 1947.

64. *Guardian*, 28 August 1947.

65. *Guardian*, 4 January 1945.

66. *Guardian*, 20 March 1947.

67. Karis, Vol. 2, *Hope and Challenge*, 117.

68. "Manifesto," Call to Attend the People's Assembly, n.d., in Karis, Vol. 2, *Hope and Challenge*, 398.

69. "The People's Charter," adopted at the People's Assembly for Votes for All, n.d., in Karis, Vol. 2, *Hope and Challenge*, 399.

70. Ibid., 117.

71. *Guardian*, 16 September 1948.

72. *Guardian*, 30 June 1949.

73. *Guardian*, 20 October 1949.

74. *Guardian*, 9 February 1950.

75. Communist Party of South Africa, "Nationalism and the Class Struggle," *Central*

Committee Report to the National Conference of the Communist Party, Johannesburg, 6-8 January 1950.

4

Transition: From the Springbok Legion to the Congress of Democrats

That there would be a formal place set for whites at the table of the national liberation movement was never a foregone conclusion. Particularly so in the late 1940s and early 1950s. White supporters of black liberation had been most visible, if commensurately controversial, in the Communist Party of South Africa (CPSA). In 1950 the party dissolved itself rather than wait to be banned under the Suppression of Communism Act. But after the party's banning, whites who had lived their commitment to economic and social justice in South Africa through the party had few other public avenues in which to do so.

The Springbok Legion was a South African servicemen's and ex-servicemen's organization chartered at the beginning of World War II, and dissolving itself after the founding of the Congress of Democrats (COD). The Legion acted as a kind of organizational bridge for radical whites between the height of the party's popularity and the rise of multi-racial organization through the Congress Alliance, in which the Congress of Democrats was the white partner organization. The Springbok Legion was chartered as a non-racial organization, although in practice it was run by white men. It was considered a progressive organization by its friends and a communist front by its foes. In fact, many members left the organization because they were alienated by the Legion's more radical participants who sought to balance the Legion's chartered goals, the promotion of ex-service people's rights and needs, with their own more far-reaching and political agenda. The inability of these more radical Legion members to co-exist with their soldierly peers pointed to the insecurity radical whites faced when seeking participation in politics and anti-apartheid work. Whites who sympathized with African goals were not necessarily trusted by politically involved Africans. They were probably trusted even less by other whites who found themselves working with white radicals in organizations like the Springbok Legion.

The influence of World War II and the generally liberalized rhetoric that pervaded allied discourse gave hope to blacks in South Africa and colonized people generally that the liberty hard fought for in the war would be extended to all people. For some, there was a sense during the war that South Africa's political culture was changing. There was hope that the tolerance exhibited by government and elements of the public toward the Communist Party, and that liberal democratic notions pertaining to rights and race, might be carried over into post-war society. South Africa might be saved from fascism at the same time the battle raged in Europe and the Pacific. These hopes were misplaced, and the window of tolerance that communists and others working at South Africa's periphery had enjoyed was closed, quickly and for good.

Thus, at a time when African nationalism was jelling and inter-ethnic cooperation was on the rise, whites had no organization through which to participate, if so invited. The leadership of the ANC was aware of this and was concerned by the lack of white participation at what seemed to be a crucial moment in the genesis of anti-apartheid organization, specifically the Defiance Campaign. Whites were thus invited to join a fledgling multi-racial alliance, which would become the backbone of anti-apartheid organization in the 1950s. This decision and subsequent invitation illustrate how greatly changed the outlook of a number of former ANC Youth Leaguers was, men who had formerly spurned white involvement in a black struggle. Their decision to call whites to action was, in essence, a distinctive tactical and ideological choice that not only had ramifications for future anti-apartheid organization but foreshadowed the divisiveness within the African camp spurred by the presence of whites. That division led to African nationalist fracture.

The Springbok Legion Through the War

Entry into the war ushered in a "fluid period" in South Africa, generating expectations of government reform and heightened respect for human rights. In early 1942 Prime Minister Jan Smuts declared that "segregation had fallen on evil days."[1] Government wartime reforms, such as the extension of feeding programs, pension schemes, and invalid aid to blacks, fueled these hopes.[2]

The Army itself sought to inculcate a more open and democratic frame of mind among white volunteers. The Army Education Service (AES) and its counterpart in the Women's Auxiliary Air Force lectured the troops on a wide variety of subjects including socialism and communism, trade unionism and 'Native,' Indian, and Coloured affairs. The war was presented as a "struggle for a better world, for democracy, for human rights." The thrust of the lecture programs was to "inculcate 'a liberal, tolerant frame of mind,' "[3] no small charge given the audience involved.

The Springbok Legion emerged out of this atmosphere in December 1941. Three organizations joined to form the Legion: the initial Springbok Legion formed by members of the Ninth Recce Battalion of the South African Tank Corps at Kaffirskraal; the Soldiers' Interests Committee formed by members of the First South African Brigade in Addis Ababa; and the Union of Soldiers formed by members of the same brigade in Egypt.[4] Fred Carneson was one of the founders of the Union of Soldiers. Carneson, having left South Africa in 1939, spent most of the next five years in Somaliland, Abyssinia, the Libyan desert, and Italy. Carneson noted that after the Union of Soldiers joined with the Springbok Legion in South Africa, the Legion "became a vehicle in the South African Army for a lot of progressive thinking, on the race issue as well, amongst white South African soldiers."[5]

Shortly after its inauguration, the Legion enunciated its goals in its "Soldier's Manifesto." The Manifesto made a number of economic demands on behalf of all ex-service people. In a more controversial vein, the Manifesto called upon members to realize in civilian life the "unity and cooperation among races" that had formed on the battlefield. The Manifesto also pledged to oppose any entity that sought to undermine democracy, and, foreshadowing future alliances, pledged to support any individual, party, or movement "working for a society based on the principles of Liberty, Equality and Fraternity."[6] The Legion was open to all soldiers regardless of race or sex. The Legion was also set apart from other service organizations by virtue of its stated willingness "to take political action in accord with the principles and practices of democracy."[7]

In the wartime atmosphere one might expect acceptance among armed forces personnel of these general and somewhat ambiguous formulations. And the pro-democratic sentiments evinced in the Manifesto could be interpreted in such a broad manner as to suggest the Legion's desire for continued (white) parliamentary government while leaving the relationship of blacks to democracy undefined. Still, the Legion leadership was immediately aware that in opening the Legion to all races controversy and criticisms would ensue:

> The Springbok Legion has always opened its ranks to all serving South Africans of every colour and creed. A considerable number of Indians, Malay, Coloured, and African soldiers have already applied for and have been accepted for post-war membership of the Springbok Legion, and no distinction has been made between them and their European comrades under arms. Springbok Legion leadership is fully aware of the fact that the organization will stand or fall on this question, and is not afraid of facing the issue.[8]

In fact, while the pronouncements of some Legion members on issues pertaining to the rights of blacks generally in South Africa proved controversial, the fact that the organization was open to African and other black ex-combatants appears to have been much less controversial. The distinctions that the Legion's more radical members failed to make in pressing claims for basic civil and human rights for white and black soldiers and civilians alike were the source of divisiveness in Legion ranks and, as South Africans' memories of the war diminished, would make the Legion a suspect, fringe organization.

The Legion at War

The Legion's membership in 1944 approached 60,000.[9] The number of blacks in the Legion during peak membership is unclear. A 1945 conference report noted eight established branches and two areas, including Johannesburg, Benoni, Durban, Krugersdorp, Port Elizabeth, East London, Germiston, Pietermaritzburg, Cape Town, and Pretoria. There were also forty-nine camp committees in the Middle East and Italy. The October 1944 issue of the Legion's publication *Fighting Talk* noted that ninety-eight new black members were enrolled in September and the Legion's employment bureau had found work for sixty-three returned black soldiers. It was strict South African Army policy that no black soldier be allowed to operate a ballistic weapon. Black soldiers were often found in front line areas working as drivers, stretcher bearers, porters, and in other occupations with no means of defending themselves save spears. Blacks would not be given the opportunity to kill whites, even in the guise of the enemy. Black soldiers were particularly vocal in their disgust at this policy, which put them at the front line without weapons. One Coloured Cape Corps member wrote to voice his disgust at government weapons policy and his treatment generally:

> Why doesn't South Africa get down to it and fight a total war? We are still unarmed -- yet you find white soldiers 500 miles beyond the front lines counted as combatants; while men of the Cape [Coloured] Corps, the I.M.C. [Indian Military Corps] and the N.M.C. [Native Military Corps] are sometimes right in the front lines without guns to defend themselves. We feel the colour bar most when we get leave for a few days. From the day we get into the rest camp until the day we leave again for the lines, the various rules and regulations seem drawn up so that we cannot forget for a moment that we are Coloured.[10]

The Legion, while vociferous in its call to arm black combatants, framed this demand in a patriotic cloak: A larger *armed* force would provide greater security for the home country.

As active duty soldiers were not allowed to participate in non-military organizations, the Legion had organized a sympathetic civilian alter-ego, the Home Front League, to act in its interests for the duration of the war. To promote demands for rights for black soldiers, the Home Front League established the Non-European Soldiers' Dependants League, which lobbied for increased allowances for black soldiers' families.[11]

In 1944 the Legion held its first national conference and adopted the Services Security Code, a statement of principle and a program that clarified the definition of veteran as including "the Coloured, the Indian and the Native ex-servicemen, for we consider that they have served their country as well as anyone else and must be provided for with the same adequacy."[12] The Code suggested that the goals of the Legion for the post-war period were largely welfare-oriented, although the Legion was willing to use political means to attain them.

The government's demobilization plan, quite generous for white soldiers and comparatively paltry for their African counterparts, was the type of issue the Legion took up. While the Legion was allowed to nominate representatives to local demobilization committees, government policy lay outside their grasp. As the war neared its end, the Legion's position was decidedly ambiguous. The Legion was recognized by, and served in, a number of capacities at the pleasure of the government.

Yet at the same time the Legion enunciated a strong negative critique of the government, its prosecution of the war effort, its demobilization program, and its general treatment of black South Africans. The radical positions underlying these negative critiques made the Legion an improbable bedfellow for its more staid soldierly brother organizations.

The Legion expended substantial energy attempting to ally with the British Empire Service League (BESL). The BESL finally acceded to the Legion's request for a meeting, and at that meeting demanded that the Legion delete the "political action" clause from its constitution. The Legion retorted that their organization would stand up for the right of returning soldiers to use political means to fight fascism in South Africa.

The meeting, and all hopes for unity among service organizations, ended when the spokesmen for the BESL stated that cooperation with the Legion was impossible, and that the Legion was a "pressure group, possibly a political party" that should "sail under its true colours and change its name to the Soldiers' Communist Party."[13]

In fact, the Legion's leadership did include strong voices from the left, among them Bram Fischer, a lawyer and Communist Party member, Cecil Williams, a playwright and Communist Party member, and Rusty Bernstein and Percy Jack Hodgson, also Communist Party members. However, there was also a highly visible liberal presence at the leadership level that did not share the radical critique. Jock Isacowitz was a Legion founder and one of the Legion's most prominent liberals. He also served on the Government Committee for Rehabilitation of Soldiers and was involved in the Liberal Party's formation in 1953.[14]

The Legion: Among Soldiers

Legion members who sought to proselytize their fellows while in Africa and Europe had a captive audience for the political work they undertook. The Legion had the attention of thousands of soldiers, soldiers fighting German and Italian fascists while at home Afrikaner anti-war elements openly expressed their solidarity with the enemy. As Fred Carneson, a Legion member, explains, the Legion had only to make the equation between the two:

> When you look at the problem of bringing the whites across as an abstract theoretical thing you can see all sorts of bloody problems. But given the right climate, as existed, for instance, amongst soldiers during the war, we were able to do an enormous amount of work in a progressive direction amongst them. We took up all sorts of issues there -- not only the question of increasing family allowances and things that were hitting their pockets and their families but political issues calling for sterner measures against the Broederbond and the Ossewabrandwag....the bulk of the South African Army were Afrikaners, not English speaking, and they also were bloody fed up with this lot. Some of them were being beaten when they went to their home towns and dorps by these anti-war elements.[15]

The anti-fascist theme was a focus of the Springbok Legion's work among white soldiers. Legion printed material linked battlefield themes and the Nationalist and ultra-Nationalist threat at home. Fred Carneson, who later edited the Communist Party's paper, *New Age*, edited two papers in the field during the war, one an official Army paper, the other a more progressive party war newspaper.

Legion members engaged in more direct forms of proselytization as well. Legion members made a conscious effort to attend the meetings called by the AES on specific topics, and there would raise issues of concern to service people, including African service people, putting forward a progressive point-of-view.

A recurrent theme in the Legion's political work is that of speaking for Africans to a white audience. Much rarer were occasions when Legion members enjoyed the opportunity of speaking directly to black soldiers. Wolf Kodesh volunteered for the Army in 1940 and joined the Legion while participating in the Abyssinian campaign, later returning to fight the Nazis in Libya. Prior to engaging the enemy in the second northern campaign, Kodesh's company and a large company of Africans and Indians took part in a training exercise at which the white soldiers had guns, the blacks, spears.[16] One evening Kodesh overheard an Afrikaner corporal giving a lecture as an assigned duty to the African soldiers. "[He] was lecturing to them and he was sick, you know, an Afrikaner you see, it went against his grain. So I said, 'listen man, you don't want to have to lecture them, do you? I'll lecture them.' "

Kodesh sent the Afrikaner corporal away, and posted guards behind a little ravine to keep an eye open for those who might be hostile to his message and its intended audience:

> I was talking to them about the Springbok Legion, and about getting equal pay and so on, and at the time it was also this whole thing about the second front. There was talk that the allies would swing the whole war around, so that they could attack the Soviets, you see.

Kodesh remembered obtaining Springbok Legion and other progressive materials in the camps without censorship. He would hold meetings in camp to discuss issues of concern to the soldiers and the Legion generally, and this rhetorical activity led to his regular transfer from one platoon to another. At the time of the particular clandestine lecture Kodesh describes above, one of the platoon leaders was a British soldier, Tuttle, who had come to Durban and joined the South African Army. Kodesh had challenged Tuttle, who had no sympathy or respect for the Soviets, on other occasions:

> So I'm telling [the African soldiers] all this. I said, don't forget, we haven't come to fight against the Russians, we have to fight against the Nazis, you have to get armed, you have to get equal pay....And as I said this I was looking at one chap, and he was looking, staring over my shoulder, and I thought, Christ. And I hear a voice saying, oh, is this where the world finds you Mr. Kodesh? And there was Tuttle.[17]

Kodesh was placed under arrest and tried, but for reasons unknown to him he was exonerated.[18] Kodesh later took part in the Italian campaign, where the juxtaposition of the suffering Italian poor and the flourishing upper class had a further radicalizing effect upon him: "I could see the connection between those poor Italians....[a]nd the blacks, and the Coloureds particularly, that I'd lived amongst, you know I thought, good Christ, it's the same thing. What's the difference? It had a hell of an effect."

The Legion's membership was ideologically diverse. There were members like Carneson and Kodesh, who stood solidly and unabashedly on the left. They used the Legion as a platform from which to extemporize on the rights not only of soldiers but of all South Africans, on the virtues of democracy and on the achievements of the Soviet Union. They defined themselves as belonging "to the workers and the common people" against other, wealthier and more influential Legion members who belonged to "the millions and the mines."[19] While some members of the public, press, and government considered the Legion to be so riddled with "reds" that it was subject to red-baiting and charges of party politicking, members like Kodesh and Carneson were probably among a small, relatively radical minority at the end of the war. This was a particularly vocal and energetic minority, however, and it was probably by virtue of their ability to mesh soldierly and more general political concerns in the war-time atmosphere that the Legion held its diverse membership together even when those members held diametrically opposite views. At the first Legion conference in 1945, a motion was put to bar blacks from the organization. The motion was defeated and another, calling for racial cooperation, was accepted.[20] Still, in 1945 the Legion leadership might well have asked how long this uneasy coalition of soldiers politically left, right, and center could be held together on the basis of veterans' claims. With war's end the Legion had choices to make as to its direction and purpose, and the choices made greatly changed the character of the organization.

The Legion in the Post-War Period

South Africa's participation in the war was costly. Nine thousand dead, thousands without jobs or homes, and a significant and growing threat from regressive Afrikaner politicians. The Legion faced a taxing dual agenda. White veterans' immediate concerns were clear: jobs, homes, and a fair dispensation from the government. Black veterans' needs were greater still. It would soon become clear to radical elements within the Legion leadership that it was incapable of pushing a liberatory political agenda while maintaining its role within the broader white South African social, economic, and political formation. If the Legion was going to lead the assault against the Nationalists, it had to appear both sufficiently democratic to remain true to its own principles and sufficiently moderate to appeal to Afrikaners. The Legion could not afford to raise the specter of basic structural changes.

Legion activities included agitation for more housing. Legion members occupied unoccupied houses and apartments.[21] The Legion continued to press for the interests of black ex-servicemen. A Legion official, discussing the government's demobilization plan for Africans, suggested that "whereas the views of the European soldier have constantly been taken into account, almost no attempt has been made to find out the hopes and ambitions of the African soldiers."[22] There was very little the Legion could do to meet African expectations. As M.N. Makobane put it, African youth had waited for a better Africa: "Unfortunately it has become a worse Africa."[23]

The Legion's transformation from a veterans' welfare organization into a more broadly political organization with an increasingly radical orientation was linked to a number of external political changes and internal struggles. In hindsight Legion leaders attributed it in part to the fact that by 1946 or 1947 the white ex-soldier's demands were largely satisfied.[24] By 1947 the Legion's initial purpose ceased to exist. While the same could not be said for black ex-servicemen's needs, by 1947 the Legion had "learned the bitter lesson that Non-European ex-soldier demands were incapable of solution by an ex-service organization alone. Only changes in the economic, social and political structure of the country as a whole could bring any relief. As a consequence the Legion had already directed its Non-European membership into their appropriate National Liberatory Organizations."[25] As Legion radicals took more vociferous positions, Legion defections increased. When the Legion leadership criticized the government's refusal to issue passports to two Indian leaders, discussion of the leadership's action resulted in the walkout of four delegates, led by a Colonel Rood, who stated that the color bar was a "sacred principle of life" to him.[26] Still, the Legion core continued its agitation, putting substantial effort toward defeating the Nationalists in the 1948 election. The Nationalist victory spelled the end of the Legion's "welfare" period, with a few exceptions. The Legion turned to political responses to Nationalist legislation. At the time of the passage of the Suppression of Communism Act politically conscious Legionnaires who did not consciously locate themselves on the left began to leave the organization. The combined effect of the Act and the broadening and deepening of the Legion leadership's commitment to "democratic" and, to some extent, internationalist politics was particularly alienating to many Legion members. This period rang the death knell for the Legion as it had been. A malleable, emotional, and conscious ex-soldier's force continued to exist through the early 1950s, but the high watermark of its concern was constitutionalism. Most Legion members accepted the Nationalists' victory in the context of white laws and a white constitution.

The Springbok Legion, the Torch Commando, and Constitutionalism

Upon winning the 1948 national election, the National Party pledged to introduce legislation that would strike Coloured voters, the last vestige of black common roll voting participation in the Union, off the voting rolls. In return, Coloureds would vote on a separate roll to elect Coloured representatives. The government attempted to circumvent the Constitution's strictures, which stated that a two-thirds majority of both houses sitting in common session was necessary to pass legislation amending the Constitution. The government had introduced its Bill in 1951 and the Bill had passed by a simple majority. The Supreme Court heard arguments in the case and nullified the legislation, a temporary victory for those bearing the brunt of Nationalist legislative initiatives.

This constitutional struggle highlights what were, in the first years of the 1950s, two antithetical strains of protest: one white, constitutionally grounded, and generally keeping within the bounds of legality; the other largely black, extra-legal, decrying the notion of constitutionality in the context of an apartheid society. Many whites within the Springbok Legion who identified with black grievances and supported black organizing efforts also found possibilities for using the anger and energies of whites who had been moved by the Nationalists' initiatives to circumvent the constitution, energies that were funneled most notably into an organization known as the Torch Commando. The Legion's numbers had diminished, as had its influence, by the early 1950s. Radical whites were attempting to work within organizations with which they shared only the barest commonality of outlooks. For these reasons the sojourn of Springbok Legion members in the Torch Commando would be brief but critical, highlighting ideological incompatibilities among politically active anti-Nationalists and propelling Legion members toward a place, as subordinates, in the black structure of protest organized around the ANC.

The Torch Commando has received little scholarly attention. Two scholars have treated the organization at greater than incidental length, and their interpretations of the Legion's role in the Commando vary considerably.[27] The first interpretation, that of Gwendolen Carter, suggests that the Commando arose on an *ad hoc* basis. A handful of ex-servicemen organized the public meetings of May 4, 1951 which brought out thousands in Port Elizabeth and Johannesburg. The assembled resolved to protest the government's abrogation of the constitution and to force, by constitutional means, an immediate general election.[28] The Torch Commando (initially known as the War Veterans Action Committee) was formed days later and planned the "steel commando drive," a vehicular cavalcade from Johannesburg to Cape Town that would present a petition to the government demanding respect for the Constitution. In this interpretation Legion members were early supporters, but not co-founders, of the Torch Commando.

The second interpretation, that of Michael Fridjohn, differs substantially from Carter's, placing the Legion in a relationship with the United Party (UP), out of which the Torch Commando was born. Here, Fridjohn traces the Legion's relationship to the UP to 1948, at which time the Legion had called upon its members to vote for the UP. In subsequent discussions Legion members had met with former Prime Minister and UP leader Jan Smuts. The two parties reached a consensus on the need to force an election and defeat the Nationalists prior to the impending reapportionment of parliamentary seats or the inclusion of South West African representatives in parliament:[29]

> General Smuts agreed with us -- as he had done on previous occasions -- that the steps taken by the Nationalists to entrench themselves would one day call for a mighty extra-parliamentary struggle, even to the point of calling for a standstill of commerce and industry in order to crystallize the demand for a general election. General Smuts also agreed with us that the attempt to remove the Coloured voters from the common roll would be the most suitable issue for a wide-scale mobilization of the opposition. He emphasised, too, that such a campaign would have to be under way before the setting up of the Delimitation Commission.[30]

If the Nationalists could not be forced to call an election before the reapportionment, it was conceivable that they would be unbeatable in future elections.

Fridjohn notes that the Legion-United Party relationship had experienced a lull with Smuts' death but was rekindled by early 1951, at which time a mutual decision was taken to mobilize disgruntled veterans in support of an early election. This meeting was allegedly attended by Cecil Williams and Jack Hodgson, Springbok Legion officials and key radical Legion members. Fridjohn suggests that the Legion and UP colluded in the formation of the Torch Commando because both felt that public knowledge of the Legion-UP partnership would have proved disastrous. The Torch Commando, a new non-political ex-servicemen's organization, would have a public leadership of unimpeachable character.[31]

The "Steel Commando" was to be the Torch Commando's central effort. In a report after the fact, the Legion's chairman noted that it was the Legion that took the initiative, even sending its General Secretary to Cape Town to prod the UP and Labour Party to action.[32] It then appears that the United Party's leader, Strauss, concerned about the UP's proximity to these potentially controversial activities, backed off from further UP participation.[33]

The Torch Commando's Steel Commando mission to Cape Town went awry as a result of UP indecisiveness. The Commando had planned to demand that the government call an election upon its arrival in Cape Town. The second interpretation states that UP leaders flew to Cape Town to implore the Commando leadership not to demand the government's resignation, as the UP was unprepared for a national election. On that night, May 28, 1951, after the official meeting, a crowd marched on Parliament, and 160 people were injured during a clash with police.[34] Fridjohn suggests that the UP's incompetence led to agitation provoked by Legion members who had "little to lose on a gamble of insurrection" and that the agitation in turn led to the marginalization of Springbok Legion members within the Torch Commando.[35] The Legion thus claimed responsibility for the formation of the Torch Commando and for infusing the organization with the will to action through Legion leadership. Legion documents acknowledged that attacks on the Torch Commando, which charged that it was controlled by the "Communist Springbok Legion," led to the ouster of Legion "activists" from the Commando.

By April of 1952 the Legion leadership was speaking of its relationship with the Commando in the past tense. The chairman's report discussed internal dissension surrounding the Legion's relationship with the Torch Commando. Some Legion members preferred that the Legion have no relationship with the Torch Commando, deeming it timid and reactionary. The Legion chairman himself noted that the Commando "showed the prejudices and reactionary attitudes of White South Africans [generally]."[36] If the Legion was responsible for the Torch Commando's formation, the creator found itself cast out by its own creation. The question then must be: What had the Legion hoped to accomplish through the Commando? Perhaps its actions were fueled by a naïve if understandable hope that the anger of white ex-servicemen could force the government's hand and at least open up the playing field for one more round of political football, or even lead to the government's overthrow and the succession of a more ideologically acceptable regime.

At a time when the Legion's influence was at its nadir, the Commando may have appeared to offer the Legion renewed influence by virtue of the Commando's linkage with the UP, an organization the Legion's radical leadership would not otherwise have consorted with. If Legion members were actually cast out of the Torch Commando, as communists and agitators, this served to underscore the quandary of the Legion's most vociferous members. These whites had watched the activist momentum shift, from the Communist Party, silenced by banning, to the African National Congress, energized by new leadership, new ideas, and the seriousness of the situation on the ground. They had led a once powerful Springbok Legion that was now a shell stripped of its more populist membership and marginalized as a communist front. Legion members embraced the signs of increasing black radicalization while working with white constitutionalists, and they must have known that they were caught in the breach, homeless, linked to no force or movement that offered any real hope for a greatly changed, less racist, more socialist future. In the meantime the Legion was rife with dissension about the Legion's relationship to the Torch Commando. The Legion chairman described this period as the "worst phase of [the Legion's] life."

Defiance and the Formation of the Congress of Democrats

As the Legion was distanced from the Torch Commando, the Commando fell in upon itself, unable to bear the weight of its contradictions. The United Party attempted to prove UP orthodoxy in Nationalist supporters' eyes by supporting government initiatives that closed off avenues of protest, so setting the stage for the party's own demise. This was not news to Legion activists who, by mid-1952, believed that the critical mass of protest was squarely with the "Non-European National Liberatory Movement":

> This movement consisting of the Indian National Congress, the African National Congress and the Franchise Action Committee [sic] represents the Indian people, the African people and the Coloured people....The method of struggle, passive resistance, is an indication of weakness, but the potential of the movement is greater than that of any other group and in the long run the political initiative will inevitably pass to it.[37]

The centripetal forces among black organizations at this time were manifest. The panoply of Nationalist legislation, including Group Areas, the Suppression of Communism Act, Bantu Education, passes, and stock limitation affected all blacks. The nascent alliance of black organizations was built upon these issues and the events stemming from them, the Defend Free Speech Convention, the May Day stay-at-home, and protests around the abrogation of the Constitution. The Franchise Action Council (FRAC) was formed in Cape Town in January 1951, with African, Indian, Coloured, and white members fighting the loss of the Coloured vote.[38] The FRAC leadership, in sharp distinction to Torch Commando counsel, spoke of " 'united action by national organizations with a view to nationwide resistance against the entire apartheid system.' "[39]

In June 1951 the ANC executive called a conference with SAIC, APO, and FRAC to discuss the general prospects for joint anti-apartheid activity. From this meeting a Joint Planning Council was created to coordinate defiance activities.[40] The basis for a formal alliance had been set.

The Defiance Campaign

In 1952, the ANC and its partners in the Indian and Coloured communities initiated a Campaign for the Defiance of Unjust Laws. Defiance campaign supporters were asked to undertake acts that contravened apartheid laws, such as standing in whites-only lines, sitting in rail carriages reserved for whites, or other similar illegal actions. "Defiers" were to court arrest in this manner, acting with complete non-violence. It was hoped that the courts and jails could be filled, making the further arrest and prosecution of apartheid law contravenors untenable. The protesters hoped by these actions to draw a great deal of national and international attention to their plight. The Defiance Campaign was unique in scale and duration. The campaign's great impact was a function of the great numbers willing to defy "unjust laws" and the extended length of the campaign, the responsibility volunteers brought to the task, and its psychological impact both on oppressors and oppressed. Over 8,000 volunteers defied apartheid laws during the approximate six month campaign. The ANC's membership increased steeply, with estimates ranging as high as 100,000 by the end of 1952, whereas prior to the Defiance Campaign estimates ranged as low as 7,000.[41]

Ruth First, reflecting on the Campaign just months after its end, stated that it had "changed the whole political face of South Africa," with blacks "intruding into the consciousness of the European population" for the first time.[42] From the standpoint of the nascent national liberation movement, that intrusion had both positive and negative aspects. The Defiance Campaign, a highly conscious challenge to the state and the status quo, was met by the state with violence, with the bannings of key leaders, and with legislation designed to stem further demonstrations. On a more positive note, the Campaign posed a challenge to whites who supported the substance of the Campaign but were not themselves involved. Monty Naicker, an important Indian Congress official, linked the importance of the Defiance Campaign to its having "laid a solid foundation for a united democratic front and it was now our task to make that a broad front of all peoples who stand for human dignity and democracy."[43]

At the time of the Campaign, white radicals were not constituted formally as a part of that front. In fact, whites were more conspicuous by their absence than their presence. There were a number of white defiers and whites who played other roles in the Defiance Campaign, but that number was small. The paucity of whites willing to defy was a matter of concern to the African leadership of the budding mass movement.

Whites in the Defiance Campaign

Dr. Moroka, President of the ANC, declared at the time of the Campaign's initiation that blacks were "prepared to work with any white man who accepted the principle of equality."[44] Moroka and other Congress leaders would have to wait until the Campaign was nearly over before they would be joined in acts of defiance by whites. Whites participated in the Defiance Campaign in a number of guises, including the provision of cars to drive defiers to more remote defiance sites, as described by Wolf Kodesh:

> While I'm driving, they'd ask me questions, I would ask them questions...."what do we do in a situation like this," I would say, well, what do you think and we'd work it out. But what everyone knew, me, as a white, and them, is you didn't hit back. You just go, and you defy, and you go to jail.[45]

Very few whites actually went to jail for acts of defiance. The single most highly publicized act of defiance by a white was that of Patrick Duncan, son of a former South African governor-general, who resigned his colonial service post in Basutoland to defy on December 8, 1952. Duncan was joined that day by six other whites who constituted a rather eclectic group, including Freda Troup; Bettie du Toit, the Afrikaner union organizer; Dr. Percy Cohen, a dentist and former member of the Communist Party; Selma Stamelman, an anthropologist; and Witwatersrand students Margaret Holt and Sid Shall. In Cape Town, four whites, Albie Sachs, Mary Butcher, H. Rochman, and A. Harrison, joined Africans in defying on the same day at the Central Post Office.[46] Bettie du Toit, more than many whites who had expressed solidarity with Africans, had paid the price for her activism, enduring jail and physical violence. She became acutely aware of the importance of the Defiance Campaign while in Port Elizabeth on union work:

> I remember going down to Port Elizabeth to do some trade union work, and I heard Africans singing, and I got up from my hotel bed to go look, and I saw Africans that were going to defy, and their heads were held so high, and I could feel they felt they were human beings, I've never forgotten that, the way they looked, the pride on their faces.[47]

When du Toit returned to Johannesburg, she attended a meeting of whites that had been called to discuss the Defiance Campaign. Du Toit spoke at the meeting, chastising whites for "never really get[ting] involved in the struggle with the African people," for always doing the "nice things." She suggested that those gathered at the meeting defy. Her suggestion was not met with approval. However, shortly after the meeting she was approached by Maulvi Cachalia of the Indian Congress, who told du Toit that the African and Indian Congresses wanted to show that there were whites who solidly supported the Campaign:

> Then Maulvi told me that Patrick Duncan had approached him, and that Patrick was going to defy, he didn't care whether anyone else was going. So I said, well, I've got these people who would defy with us, so he said fine. Patrick was very pleased, so we defied an unjust law....They arrested us and took us into prison, we spent the weekend there, the whole lot of us.[48]

Du Toit and Troup ended up spending six weeks in prison, an unusually stiff sentence undoubtedly set to make an example of them. Commenting shortly after their acts of defiance, the whites spoke with a mixture of resolve, anger and foreboding. Troup spoke of the possibility that the racial divide would "widen beyond all bridging" and suggested that the least white South Africans could do would involve engaging in acts of good faith until more reasonable leaders acknowledged the equality of all South Africans.[49] Stamelman, the anthropologist, stated that she could no longer find it in her conscience to be "even a possible party to repression."[50]

White Cape Town volunteers, who had worn ANC armbands while defying, were greeted with a rally at the Drill Hall attended by 500 supporters. Albie Sachs pledged his support to Congress and noted that South Africa was his homeland. He vowed to do all in his power to make the country a home for all South Africans.[51] Mary Butcher, speaking with great vehemence, admonished whites to wake up and "stop swallowing the filthy propaganda of race hatred vomited forth by Malan and his supporters." She called on whites to set aside "selfish prejudice" and support the majority "in their legitimate demands for equality and justice."[52] If these whites' actions, given the numbers involved, appeared symbolic, they were also important. Their actions constituted proof that some whites were willing to make a commitment to the creation of a just society. At a time when white activism was in abeyance, the actions of the defiers bolstered the black leaders who believed that there was a place for whites in the national liberation movement.

The only organized group of whites that openly and actively supported the defiers was the now much diminished Springbok Legion. At the time of the Defiance Campaign the Legion, no longer a presence within the Torch Commando, continued to push its plan for a national work stoppage. When the Defiance Campaign was tabled, the Legion issued a statement of support declaring that "this movement had set on foot a great political upsurge of Non-European democratic feelings, which should meet with the fullest support of all anti-Nationalist Europeans."[53]

Among the Legion's contributions to the Campaign was the pamphlet, " 'D' Day for Democracy." The pamphlet was unequivocal in its support for the Campaign, and in its placing of blame on the government and white community. The pamphlet's author proffered two visions of the South African future. One was steeped in increasing bitterness, hatred, and "the dark presence of constant and recurring clashes leading to a bloody and destructive end." The "other road" foresaw a white role in speaking up against injustice and taking an active part in all anti-Nationalist organizations to fight for the recognition of the Defiance Campaign. This would prove to blacks that whites sought "justice for men of all races." The Legion author envisioned that from this activity "can flow a new golden era for us all....an era in which fear and race hostility will give way to friendship and peace."[54]

The Springbok Legion was becoming an anachronism. The war that had led to its formation was long over and had given way to a new battle, for which the Legion, carrying its ex-service baggage, was not chartered. Yet the Legion was the closest entity to a white brother organization of the African, Indian, and Coloured Congresses then in existence. It had stood with them in various capacities and supported them in the new decade. The impetus for the creation of such a brother organization would come from the African and Indian Congresses, and the Legion would play a key role in its formation, providing a certain continuity between the activism of the war and the activism of a newly organized and energized black population.

The ANC and the Call to White Democrats

In November of 1952 Oliver Tambo, Walter Sisulu, and Yusuf Cachalia, representatives of the African and Indian Congresses, called a meeting in Johannesburg's Darragh Hall of 200 whites who had supported the Defiance Campaign.[55] The group at the hall was ideologically diverse, including both liberals and avowed communists. The ANC was compelled by both positive and negative factors to call the Darragh Hall meeting. A small group of white South Africans had shown a willingness to make sacrifices for the Campaign. A larger if less committed group had evinced sympathy and provided support. In a negative vein, there was the fear that, given the paucity of visible white participation in the fight from the time of the Communist Party's banning, the national liberation movement could take on clear and destructive racial overtones if the opportunity were not seized to bring whites in as partners on an active basis. At the meeting, Tambo and Sisulu told the gathered whites:

> The silence of European democrats to the challenge of the issues involved in the Defiance Campaign is being construed by Non-Europeans as acquiescence in and approval of the Government's policies. Thus rapidly creating the belief among large numbers of Non-Europeans that all whites are hostile to them and their aspirations and that the situation is being transformed into a White vs. Non-White struggle.[56]

The role whites had played in the Defiance Campaign had been the catalyst in the ANC's decision to call the Darragh Hall meeting. The minimal white presence, while suggesting the possibility of greater participation, was also cause for concern, a concern the ANC wanted to share with whites themselves. Walter Sisulu reflected on the Congresses' dilemma revolving around the question of the white role in unfolding events.[57] Sisulu noted that within the ANC "the question was what do you do with these people." The ANC discussed the matter with SAIC and decided that the two organizations should take the initiative in forming an "alliance of Europeans."

The nature of any new white organization's membership was not the only controversy to arise in this initial contact with Congress. Those members of the Darragh Hall audience who would provide the membership of the Congress of Democrats demanded universal adult suffrage. Sisulu had himself replied to the query of someone on the floor that he did not believe an offer of a qualified franchise would hold "any appeal" for most blacks. The banned Communist Party and the Springbok Legion were well represented at the Darragh Hall meeting. These were the two organizations that had provided homes for white democrats in the last tumultuous decade. The CPSA had been banned; the Legion, marginalized. Therefore, it was not surprising that whites who had fought to maintain organizations providing them a place in the liberation movement were present and enthusiastic. However, most of those liberals present at the Darragh Hall meeting were not prepared to participate in the formation of an organization with known communists.[58] The outcome of the meeting was the formation of an organization first known as the South African People's Congress, which shortly thereafter became the Johannesburg Congress of Democrats.[59] The organization was pledged to expose and publicize the "evils of colour bars"; to mobilize support for the abolition of discriminatory laws; to speak for equal political and economic rights and opportunities for all South Africans; and to fight for freedoms of speech, assembly, movement, and organization.[60]

The Darragh Hall meeting was an extremely important moment in the history of both the African National Congress and the national liberation movement generally. Through the first years of the 1950s the ideological orientation of the ANC was ambiguous and seemingly capable of transformation. The "Young Turks" who rose with the new self-conscious African nationalism of Anton Lembede and Peter Mda from the early mid-1940s were very much predisposed against communists, "foreign influences," and whites who might attempt to guide or derail the struggle in the guise of paternalist liberalism or leftist internationalism. Yet these were the same individuals who, by 1952, were making policy for the ANC. And the policy they devised in late 1952 to bring whites into direct dialogue with them set a precedent that has influenced the ANC's policy to the present day.

Was the Darragh Hall meeting a logical extension of the principle of unity embodied in the alliances forged with the leaders of the Indian and Coloured Congresses from the mid-1940s? If Africans were acutely aware of bearing the brunt of oppressive policies both before and after 1948, they were also aware of the suffering, historical and immediate, of the Coloured and Indian peoples. These groups could be understood as fellow oppressed: their disabilities were manifest, and there was a history of common struggle. White democrats were anomalous. Could one say they were oppressed? How did they define for themselves the reason for their participation in the multi-racial alliance? Could they be trusted to understand and accept their subordinate role therein? Could they be trusted at all? It is significant that answers to these questions, and affirmative ones at that, were provided by the ANC's Youth League at the time of the Johannesburg Congress of Democrats' formation. The Youth League authors found in the Congress of Democrats' formation a sign of the ANC's growing strength:

> This [COD's formation] is a most significant step in the political history of this country. The Defiance Campaign can be regarded as the effective and proximate cause of the establishment of the Congress of Democrats. This is, in itself, a pointer to future developments. As the leaders of the ANC have often pointed out to their followers, the stronger and more powerful the ANC, the more it will precipitate from the ranks of the Europeans and indeed other groups, genuine democrats.[61]

The author noted that the idea of "liberation" necessarily had quite different connotations for whites than for Africans. When the African spoke of liberation, he or she tended to think in terms of the ending of caste or national oppression. Liberation had a different meaning for the white democrat. For the white democrat, "liberation" connoted:

> Freedom from class oppression. In other words every time a European in this country speaks of liberation his mind must jump to the concept of freedom from economic exploitation under capitalism. Being relatively free from national oppression this concept is the one that occurs immediately and

> spontaneously to him....This crudely put is the contradiction that has always existed as between the European and African democrats in this country.[62]

The Youth League author was suggesting that radical whites would have no natural affinity for the concept of national struggle. The white democratic eye viewed every situation through Marxist spectacles, which provided a narrow class view. This was a highly perceptive comment on the part of the Youth League author, pinpointing as it did the challenge facing white radicals who stood at the edge of this multi-racial national experiment. Radical whites, those in the Communist Party and Springbok Legion, had long analyzed the South African situation in class terms. In the process of grappling with the intensification of racially based law and the maturation of a new and vigorous African nationalism, they found a new, if grudging, respect for that which they had dismissed shortly before as petit bourgeois. But now they were faced with a situation in which they were to subordinate their own analysis to the national liberation movement and its leader, the ANC, neither of which was fighting specifically for the working class or socialist revolution. How were radical whites to articulate with a movement for national liberation? In the party, radical whites were able to define their identity without reference to race, in the sense that race, while a reality, was subordinate as an explanatory tool to class forces. This was arguably more than a matter of theory or doctrine: It was an outlook, a way of viewing the world, that allowed whites of conscience in a racist society to explain their presence in that society in a personally affirming manner. However, now, the anomalous nature of that participation was evident. Would the Youth League author prove correct in his description of the unshakable radical white view, or could radical whites redefine their identity, their sense of purpose and goals, and construct roles for themselves as African nationalists, so redefining the very idea of "nation" itself?

The Springbok Legion and the Congress of Democrats

The Springbok Legion was, by early 1953, composed of a small number of committed activists who looked to the African National Congress and the Congress Alliance generally as core elements of the evolving resistance. The Darragh Hall meeting resulted in the dichotomization of sympathetic white forces. Whites who were anti-Nationalist but could not accept the ANC's call for immediate universal suffrage or were concerned with the presence of communists, particularly white communists, in the Congress Alliance, had formed the Liberal Party. The Liberal Party was the outcome of the growth of a number of organizations that had arisen in the face of dissatisfaction with the United Party's racial policy in the first years of the 1950s. Although centered in the Cape Town Liberal Group, other groups formed in Johannesburg,

Pietermaritzburg, and Durban.[63] These groups were united in the South African Liberal Association in January 1953. In May the Association became the South African Liberal Party. The Liberal Party's theme was equal rights for all civilized men and equal opportunities to attain civilization. The party was chartered as non-racial, and the majority of its members would be African.

The Legion was highly conscious of the formation of this national liberal body, and was highly critical of it in the context of immediate needs:

> The Liberals seek to appease the non-Europeans with a few concessions and, at the same time, not to run too far ahead of European opinion. The result is a policy acceptable to no substantial part of the population. The Liberals rely on conventional parliamentary methods and offer no part in their struggle to the disenfranchised masses, who are mobilized behind the non-European Congresses. What is urgently needed today is a body of Europeans who will ally themselves with these organizations -- a body which will not seek to bargain with or buy off the non-Europeans but which will march with them to the attainment of their legitimate democratic demands....We feel the time has now come for....[the] founding of one nation-wide body which will represent all those Europeans who believe uncompromisingly in democracy.[64]

The Congress of Democrats and the Legion were attempting, in the eyes of the authors above, to serve that purpose. However, if whites were to become full, responsible partners in alliance with the African and Indian Congresses, it was necessary to put an organization like the Congress of Democrats on a national footing. The problem had been discussed at a Legion meeting in Johannesburg, and over £100 had been raised to aid in bringing such an organization to life.[65]

The need for a focused organization increased as the Legion and COD began to work together on a host of issues of national importance. By August of 1953 the Legion and Congress were cooperating on a number of issues, including fights against specific applications of Group Areas legislation, the Western Areas removals plan, and the Native Labour Bill. By the end of August there were thirteen COD branches in Johannesburg, a branch in Benoni, and one in Durban, with an organizer in Port Elizabeth. The estimated membership was 500.[66] The Congress of Democrats was a national organization and an affiliate of the Congress Alliance in all but name by September 1953, when a conference was called, sponsored by the Congress of Democrats, the Springbok Legion and the Democratic League, for the purpose of amalgamation.[67]

National Conference

The inaugural conference of the South African Congress of Democrats took place at the Trades Hall in Kerk Street, Johannesburg on October 10 and 11

1953. Eighty-eight delegates were in attendance including members of the Springbok Legion executive, the Cape Democratic League, the Johannesburg Congress of Democrats, and ten of its Transvaal branches. There were also representatives from Durban and Port Elizabeth.[68] The conference's first resolution provided certain philosophical, moral, and political guidelines for the creation and conduct of the organization. In the first paragraph of this resolution the participants pledged their support for the "Universal Declaration of Human Rights" and repudiated all racial and apartheid doctrines. The second part of the preamble provided more of a programmatic sense of COD's intentions:

> We proclaim our conviction that racial conflict and national oppression are linked with international conflict and war which threatens the advancing standards of life and liberty of all mankind.
>
> We therefore found this association to advocate the principles of equality and the brotherhood of men, to strive for the maintenance of world peace and the ending of national discrimination and oppression, and to win South Africans to support a programme of extending rights and liberties for all our people.[69]

If the emphasis was subtle, there was nonetheless an important connection here between South Africa's "unique" problems, the world political economy, and the conduct of international relations under the world capitalist system. This outlook came naturally to the founding members of COD, many of whom were Communist Party members and union organizers steeped in proletarian internationalism.[70] The Congress of Democrats' particular interest in international issues would manifest itself a number of times throughout the 1950s. However, the gathered delegates were still acutely aware of the primary purpose for the formation of the Congress, the securing of all common and universally accepted rights for the black peoples of South Africa. Specifically, COD would attempt to "win support for our policy amongst those sections of the population not catered for by the Congress Movement." COD's charter was therefore clearly delimited to work among the white community. Elections were held for president, vice president, secretary, and twelve national executive committee seats. Piet Beyleveld and Jack Hodgson, both Springbok Legion leaders, were elected president and secretary, and Len Lee-Warden, one of those responsible for the formation of the Cape Democratic League, was elected vice president. The twelve Executive seats were filled by Johannesburg residents, where COD was to be headquartered.[71]

Rusty Bernstein, a leader of the banned Communist Party, a member of the Springbok Legion, and one of the movers behind COD's formation, gave the keynote address.[72] Bernstein's address was particularly interesting for its analysis of the critical nature of the moment. With successive NP victories and the rapid extension of racialism and "fascist reaction," paralleled by the rapid growth of the Non-European liberatory movement, events in South Africa were moving the country toward a clash, and it was incumbent upon people to "take sides." Bernstein stated that the struggle was taking on a racial appearance:

> By the fact that nowhere in the democratic camp is there a body comparable to the Congresses, composed predominantly or largely of Europeans, and campaigning predominantly amongst the European population. Nowhere is there a body on a Union-wide scale which enters into all the political and social affairs of the European population, in open disagreement with white supremacists....This void in the democratic camp must be filled, and filled soon, if the democratic camp is to challenge South African reaction not in the name of the Non-European people, but in the name of the overwhelming majority of South Africans drawn from all our racial groups.[73]

The organization Bernstein proposed would not supersede the African, Indian, or newly formed Coloured People's Congresses (CPC) in any way. Rather, the organization would establish a working unity while attempting to draw support from "wider circles" in the white community. Invoking the legacy of the new organization, Bernstein noted some optimism at the perceived growth of a more democratic and progressive outlook among white South Africans, which he put down to the earlier work of that small group, including many of the assembled, who had "planted the seed of progressive and radical outlook in South Africa which now begins to bear fruit."

The Cape Democratic League applied to merge with COD in November 1953.[74] The Springbok Legion's situation was somewhat different, having been a substantial organization that maintained a small but dedicated following and continued to publish *Fighting Talk*, one of the few publications on the left. At its Annual Conference in late 1953, the Legion resolved that its members should "be invited" to join one of the Congress Alliance affiliates, that the Legion's assets be liquidated and any surplus be dispersed in the interest of the democratic cause, and that the Legion's National Executive Committee (NEC) discuss with COD how *Fighting Talk* could best be utilized for the benefit of the national liberation movement.[75] This was the effective end of the Springbok Legion. Its name appeared on some statements and flyers with that of COD for another year, but the momentum that the Legion had found at the time of the Defiance Campaign now propelled it naturally into the work of COD and the Congress Alliance.

The time span from the formation of the Springbok Legion to that of COD was little more than a decade, yet the situation facing those struggling for rights and democracy had changed greatly. The war had redrawn maps and political alignments, and had served to sharpen the political acumen and will of the African majority. Post-war disillusionment following a chain of broken promises and the victory of the Nationalists had invigorated the black communities of South Africa.

It was only a few white South Africans who were highly conscious of the oppressive nature of South African society and sought to integrate themselves in the struggle of blacks for justice. And those whites found few outlets for their energies. One of those outlets was the Communist Party, which reached an apex of popularity during the war years and sought to present a voice counter to the oppressive status quo through electoral activity, public demonstrations, and educational initiatives. Although the party's membership was predominantly African, much of the party's leadership remained white through the time of its banning, and among those party members there was a certain skepticism about the African national movement, although mixed with a recognition that it would be the leading force of the future.

The Springbok Legion acted as an important bridge between the Communist Party of South Africa's distrust of national politics and the multi-racial alliance that would characterize the national liberation movement in the 1950s. The Legion, originally mandated as an ex-servicemen's organization, moved from a period of widespread popular recognition as the representative of ex-servicemen, black and white, to a much smaller and more radical membership with a general critique of South African society and a willingness to stand with the African majority at the time of the Defiance Campaign. Ultimately, the Legion was unsuited to the demands of mass organization across racial lines.

The Congress of Democrats was thus synthetic, bringing together in one organization former members of the CPSA and current members of the SACP, the Legion, and various trade union organizations. Unlike the CP or the Legion, which whites had formed, COD was a product of African and Indian initiative, a subordinate organization charged with explaining the principles and policies of the black majority to whites. The Alliance was faced with a difficult, perhaps contradictory, task, for at a time when all democratic forces had found a vehicle for unity in the Congress Alliance, the government was busy closing off avenues of discourse.

Notes

1. Tom Karis, *From Protest to Challenge: A Documentary History of African Politics in South Africa, 1882-1961*, Vol. 2, *Hope and Challenge, 1935-1952* (Stanford: Hoover Institution Press, 1977), 73. The most prominent example was Smuts' February 1942 utterance that "segregation [had] fallen on evil days."

2. Jack Simons and Ray Simons, *Class and Colour in South Africa, 1850-1950* (London: International Defence and Aid Fund for Southern Africa, 1983), 540.

3. Helen Joseph, *Side by Side: The Autobiography of Helen Joseph* (London: Zed Books, 1986), 27.

4. Springbok Legion (pamphlet), "Springbok Legion: The History and Policy," 1944, Pressure Groups Pamphlet Collection, Institute of Commonwealth Studies. Dates,

locations and persons involved are in some cases matters of contention. The Union of Soldiers is noted as having been formed in Egypt in the above mentioned pamphlet, while the Simonses state that it was formed in Libya. In a *Guardian* article (14 January 1943) Dan Pienaar is given credit as partially responsible for its formation. The Simonses also mention a sole name, in this case Morley-Turner (p. 540). It is agreed that Vic Clapham, cartoonist for the *Guardian*, was the leading figure in the formation of the initial Springbok Legion organization at Kaffirskraal.

5. Julie Frederikse, *The Unbreakable Thread: Non-Racialism in South Africa* (Bloomington: Indiana University Press, 1990), 43-44. Interview of Fred Carneson conducted by Julie Frederikse.

6. Springbok Legion (pamphlet), "The Springbok Legion Fights for YOU," n.d., University of Cape Town Library, BC579 B18:9.

7. Ibid.

8. *Guardian*, 28 May 1942.

9. Carter-Karis Papers, Northwestern University, "The Springbok Legion," 2:GS2:92. This document appears to have been written by a member of the Legion for the use of defense attorneys in the Treason Trial.

10. *Guardian*, 10 December 1942.

11. *Guardian*, 14 January 1943.

12. *Fighting Talk*, Vol. 3, No. 1, April 1944.

13. *Guardian*, 9 November 1944.

14. *Liberal News: Monthly Bulletin of the Transvaal Division of the Liberal Party of South Africa*, Vol. 4, No. 2, February 1962, Liberal Party Papers, Microfilm 837, Institute of Commonwealth Studies.

15. Frederikse, *The Unbreakable Thread*, 43.

16. Interview, Wolf Kodesh, London, 12 February 1987.

17. Ibid.

18. Ibid. Kodesh thinks that his case was sent to an outside board or adjudicator for a decision. Kodesh had heard after the fact that Leo Kuper was involved in the decision, accounting for the lack of a harsher penalty.

19. *Guardian*, 29 March 1945. The latter reference is specifically to Sir George Albu, a Johannesburg city official who resigned his Legion membership.

20. *Guardian*, 15 February 1945.

21. Springbok Legion, "Report of the National Chairman," 9th Annual Conference, April 1952, Carter-Karis Papers, 2:GS2:30/1.

22. *Guardian*, 16 August 1945.

23. *Guardian*, 9 January 1947.

24. Springbok Legion, 9th Annual Conference, "Report of the Chairman," 1952. It is here, as quoted above, that the Legion acknowledged that it was incapable of speaking to the needs of black ex-soldiers by itself.

25. Ibid.

26. *Guardian*, 10 April 1947.

27. Gwendolen Carter, *The Politics of Inequality: South Africa Since 1948* (New York: Frederick A. Praeger, 1958), Ch. 12, "The Torch Commando," 302-339. Michael Fridjohn, "The Torch Commando and the Politics of White Opposition: South Africa 1951-1953," 175-202, in *Papers Presented at the African Studies Seminar, University of the Witwatersrand, Institute of African Studies*, Johannesburg, 1977.

28. The 4th May resolutions were:

1) We ex-servicemen and women and other citizens assembled here protest in the strongest possible terms against the action of the present Government in proposing to violate the spirit of the of the Constitution.

2) We solemnly pledge ourselves to take every constitutional step in the interests of our country to enforce an immediate General Election.

3) We call on other ex-servicemen and women, ex-service organizations and democratic South Africans to pledge themselves to this cause.

4) We resolve that the foregoing resolutions be forwarded to the Prime Minister and the leaders of the other political parties.

29. The gerrymandering the NP was executing through the 1952 delimitation was going to provide the NP with an increased number of seats in Parliament through the overweighting or "loading" of the rural against the urban vote. In the Commando's submission to the 1952 Delimitation Commission it was noted that "under the 1948 Delimitation [the rural constituencies] gained six seats at the expense of the urban constituencies." "Memorandum Submitted to the Tenth Delimitation Commission, 1952, On Behalf of the War Veterans' Torch Commando," 8. Carter-Karis Papers, 2:ET1:45/2.

30. Springbok Legion, 9th Annual National Conference, "Report of the National Chairman."

31. Fridjohn, "The Torch Commando and the Politics of White Opposition," 178-9. Ergo the leadership of Louis Kane-Berman, Ralph Parrott, A.G. "Sailor" Malan, Major J.D. Pretorius, a member of the Provincial Council, and Doreen Dunning, former head of the WAAF.

32. Springbok Legion, 9th Annual Conference, "Chairman's Report."

33. "The Springbok Legion," n.a., n.d., Carter-Karis Papers, 2:GS2:92. This document, from the Carter-Karis collection of Treason Trial documents, appears to be a statement drafted by a Legion leader at the time of the Treason Trial, sketching out the history of the movement, including aspects of the Legion's relationship to the Commando and the United Party.

34. *Cape Times*, 30 May 1951, 1. In the article, Legion representative Malan apologized for the "disturbances" stemming from the march, including the breaking of the fence around the Groote Kerk. He also contested yet again charges that demonstrations taking place around the country were "stunts engineered by the U.P. and political agitators."

35. Fridjohn, "The Torch Commando and the Politics of White Opposition," 189. In support of this, Fridjohn cites Janet Robertson's notation of her interview with former Commando member John Lang, where Lang states how Jock Isacowitz, a leading Legion member, approached the Commando executive with a plan to march on Parliament and take control from the Nats. Janet Robertson, *Liberalism in South Africa* (Oxford: Clarendon Press, 1971), 60 ff. Yet Fridjohn also notes that in an interview he conducted with Kane-Berman, Berman admitted that "many of the rank and file at the Grand Parade had urged the leaders of the meeting to direct a march on the Assembly to 'throw out the Nationalists.' " It is difficult to take this suggestion beyond the realm of speculation.

36. Springbok Legion, 9th Annual Conference, "Chairman's Report."

37. Ibid.

38. Brian Bunting, *Moses Kotane: South African Revolutionary* (London: Inkululeko

Publications, 1975), 179. Sundra Pillay was elected chairman, Reg September, secretary. Other members included Johnson Ngwevela, Sam Kahn, Fred Carneson, and John Gomas.

39. Karis, Vol. 2, *Hope and Challenge*, 411.

40. Peter Walshe, *The Rise of African Nationalism in South Africa: The African National Congress, 1912-1952* (London: C. Hurst & Co., 1970), 402. The JPC included Moroka, Marks, Cachalia, Sisulu, and Dadoo.

41. Karis, Vol. 2, *Hope and Challenge*, 427.

42. Johannesburg Discussion Club, *Viewpoints and Perspectives*, Vol. 1, No. 1, March 1953.

43. *Advance*, 15 July 1954.

44. *Clarion*, 5 June 1952.

45. Interview, Wolf Kodesh, London, 12 February 1987.

46. *Advance*, 11 December 1952.

47. Interview, Bettie du Toit, London, 11 June 1988.

48. Ibid.

49. *Advance*, 11 December 1952.

50. Ibid.

51. Ibid.

52. Ibid.

53. Springbok Legion, "Chairman's Report to the National Conference," 1953, Carter-Karis Papers, 2:GS2:30/2.

54. Springbok Legion, " 'D' Day for Democracy," (pamphlet), 1952, SAIRR Political Documents, Microfilm 860, Institute of Commonwealth Studies.

55. *Advance*, 27 November 1952.

56. Ibid.

57. Interview, Walter Sisulu, by Mary Benson, Mary Benson Papers, School of Oriental and African Studies, [M3233].

58. See Robertson, *Liberalism*, 88-89, for a discussion of this series of events. Robertson interviewed a number of people present at the Darragh Hall meeting who would become founders and prominent members of the Liberal Party.

59. Thomas Karis and Gail M. Gerhart, *From Protest to Challenge: A Documentary History of African Politics in South Africa, 1882-1964*, Vol. 3, *Challenge and Violence, 1953-1964* (Stanford: Hoover Institute Press, 1977), 13.

60. *Advance*, 27 November 1952.

61. African National Congress Youth League, "The Congress of Democrats," *Afrika*, n.d., Popular History Trust, Harare.

62. Ibid.

63. Robertson, *Liberalism*, 191.

64. Springbok Legion, untitled form letter, 16 July 1953, Popular History Trust, Harare.

65. Springbok Legion, "Dear Friend," 15 July 1953, signed by Cecil Williams, National Chairman, Carter-Karis Papers, 2:GS2:47/2.

66. Ibid.

67. *Advance*, 24 September 1953. The meeting was attended by approximately 50 persons, "predominantly European," and a committee of fourteen was elected to office and instructed to form branches in the Western Cape.

68. South African Congress of Democrats, "Report of South African Congress of Democrats," October 1953, Carter-Karis Papers, 2:DC2:30/1. Among the delegates were four representatives of the South African Indian Congress and five representatives of the Transvaal Indian Congress. Thirty-four other delegates are listed as "by invitation" but there is no record of ANC attendance.

69. South African Congress of Democrats, "Constitution," 1953, SAIRR Political Documents, Microfilm 860, Institute of Commonwealth Studies.

70. South African Congress of Democrats, "Report of South African Congress of Democrats." A motion was unanimously adopted, arising from the discussion of papers read at the conference, stating that COD recognize "the vital and indispensable role to be played by the working class and Trade Union movement in the struggle for national liberation. Accordingly we must make a serious effort to work and disseminate South African Congress of Democrats' propaganda among the working class and Trade Union movement and at all times strive to win support of workers and Trade Unions."

71. The Executive members elected were: Joe Slovo, Ruth First, J. Baker, Edward Roux, Maulvi Cachalia, C. Feinstein, L. Baker, Michael Hathorn, Beatta Lipman, Molly Fischer, and Helen Joseph. That Cachalia was on the executive did not lead to membership of blacks in any appreciable number. This would be a point of some controversy later and will be discussed below.

72. South African Congress of Democrats, "The Road to Liberty," Inaugural Conference address by Lionel "Rusty" Bernstein, October 1953, Carter-Karis Papers, 2:DC2:30/2.

73. Ibid.

74. *Advance*, 5 November 1953.

75. Springbok Legion, "Resolution, National Conference, 1953," Carter-Karis Papers, 2:GS2:30/4.

5

White Radicals: A Collective Biographical Sketch

Of the millions of white South Africans who lived through the emplacement and magnification of legislated apartheid in the 1940s and 1950s, only a handful raised their voices, let alone their hands, against this tide. If most white South Africans could never have imagined participation in organizations such as the Black Sash and the Liberal Party, the thought of membership in the South African Communist Party (SACP) or the Congress of Democrats (COD) probably seemed otherworldly and almost beyond comprehension. As a member of the Congress of Democrats, one was *a priori* challenging the state and South African society as a whole. As participants in the Congress Alliance and subordinates to the Alliance and its leaders, the African National Congress, COD members were acknowledging their acceptance of a radical vision of the South African future. That vision, as set out in the Freedom Charter, foresaw a post-apartheid South Africa guaranteeing basic human and civil rights to all citizens and meeting the basic needs of all citizens without regard for race. Inherent in this vision was a belief in the complete illegitimacy of the existing South African society. The ANC's leadership in the Congress Alliance was a pointer to the inevitability of African leadership in South Africa as a whole. COD members were setting themselves adrift from white society, its mores and conventions. They found meaning and purpose and, perhaps, a degree of personal liberation in their identification with African aspirations and in contact with the members of the African, Indian, and Coloured congresses as politically active peers but also as black people who could be known as friends. It was this unique willingness to identify with, and to be seen to be doing so, with the black majority, that led one COD speaker to declare, "We are the symbol of heresy against the accepted creed of white supremacy, and, like heretics everywhere, we are persecuted."[1]

Who were these self-described heretical, "democratic-minded" whites, and what led them to an identification with the movement for national liberation? Were there commonalities of heritage, religion, class, or experience that were integral in shaping those few who chose paths leading from the white fold and into the ranks of the revolutionary opposition? The speaker who described himself and his fellows as "heretics" hoped that COD's ranks would swell, spurred by the realization among other whites that the South African future lay in partnership rather than mastery. In fact very few whites, a relative handful numbering in the hundreds, embraced this realization. This exclusivity raises a question at the heart of this study: Why these particular individuals? How did the white South Africans who comprise this study come to the realization that they were living in, and by their very presence colluding with, an abjectly evil society? And what, given that powerful realization, moved them to act upon it, to commit themselves to the destruction of the order in which they were reared, in effect to commit socio-cultural suicide, severing the ties that had identified them as members of the privileged racial caste? This chapter profiles a number of individuals who were members of the Communist Party, Springbok Legion, and Congress of Democrats, in an attempt to answer these questions.

The Jewish Presence

A substantial percentage of COD's membership was Jewish. Denis Goldberg, a member of the Congress of Democrats and later active in the ANC's military wing Umkhonto we Sizwe (MK), noted of his 23 years in prison:

> As a Jew in prison, you saw it all, you see, a sergeant greets in the morning, to each political prisoner, "good morning Jew, good morning Jew, good morning Jew," whether they were Jewish or not. He gets to me. I say I don't like this and I won't have it. He says, "it's a joke." And I say it's the kind of joke I don't like. So the next morning he comes down the passage and says, "good morning communist, good morning communist, good morning communist." If you're a Jew you must be a communist, and if you're a communist you must be a Jew. If you're opposed to apartheid you must be both, whether you are or not.[2]

South Africa was a fertile anti-Semitic field, and no anti-Jewish stereotype or diatribe went wanting for a lack of torch bearers. However, Jews were not the only identifiable ethnic or linguistic group among the white South African population to produce a quotient of "heretics." Of the thirty-seven individuals involved in the relevant organizations and for whom biographical data was available, twenty-five were Jewish. There were also five Anglicans, a Catholic and four Afrikaners. While the vast majority of South African Jewry did not participate in radical politics of any type, there were a small number of

Afrikaners who did. The presence of Jews in highly visible roles in organizations such as the SACP and COD fueled prevalent stereotypes of Jews, and anti-Semitism generally. This section will examine the origins of the South African Jewish population and the foundations of its political life in South Africa.

Early Jewish Settlement

The first Jews to settle at the Cape were soldiers of the Dutch East India Company who converted to Christianity and were baptized in 1669.[3] With the transfer of the Cape to the British and the subsequent decrease in religious persecution, it became possible for Jews to create a community and openly practice their religion. Many Cape Jews in the early nineteenth century were shopkeepers and tradesmen. A number of Jewish families who would gain prominence arrived with the 1820 settlers, the first substantial infusion of British settlers. By 1841 a sufficient number of practicing Jews were living at the Cape to allow for the assembly of a *minyan*, a religious quorum which must be composed of no less than ten orthodox Jewish men. In 1859 the new rabbi at the Cape, Joel Rabinowitz, surveyed the Jewish population outside Cape Town and found nearly sixty families, most from England and Germany, living in the countryside in towns such as Burghersdorp, Mossel Bay, Ceres, and Victoria West.[4] Many were traders, but one also found doctors and public officials. There was also a minor influx of Jews among those tempted by the discovery of diamond deposits at New Rush, later renamed Kimberley. Jews were to be found in all roles there, as diggers, buyers and provisioners.

The Jews who arrived before the 1880s left an imprint on the South African cultural and economic landscape. They organized Jewish cultural and religious practice, they traded, and they innovated. A few became rich, and, like mining magnate Barney Barnato, are remembered today. However the English orientation of the Jewish community was strongly challenged from the 1880s with the influx of eastern European Jews fleeing persecution or seeking a new and more comfortable life beyond the welter of Europe. From this community came many of the children who would cast off sectarian interests in favor of identification with the struggle of the black South African majority.

The Jews arriving from eastern Europe were a strikingly homogeneous group, the majority originating from a handful of Lithuanian provinces and Byelorussia.[5] The South African Jewish population, 4,000 in 1880, had expanded to just over 38,000 in 1904 and nearly 50,000 by 1911.[6] These influxes of Lithuanian and Russian Jews were fueled by a wave of pogroms initiated in southern Russia in 1881. Collections were taken in the larger Lithuanian towns and in Cape Town to aid emigration, and these efforts sparked the general exodus that led to the departure of three million Jews between 1884 and 1914.[7] The influx of eastern European Jews subsided during the first decade of the twentieth century, setting the parameters for the South African Jewish community, which would grow from that time largely through internal expansion. A study of the South African Jewish population conducted in 1935 found that among foreign-born Jews, 52.4% were born within the boundaries of Lithuania, 12.6% were from Poland, and 14.3% were from Russia.[8]

Michael Picardie's family history stands as a blueprint of much eastern European Jewish immigration. Both parents were born in Lithuania, the mother from Wilna and the father from a village near Kovno. Upon arrival in 1905, Picardie's father was six or seven, his mother one or two. Yiddish was the family language, and no one spoke a word of English. He remembered that newly arrived Jews such as his grandparents were referred to as "Peruvians"[9]:

> There was a hell of a lot of anti-Semitism against them. The first jobs that they did [were] running illicit liquor canteens, for example in Johannesburg, or my paternal grandfather ran what was called a kaffir eating house. I quote the phrase that my grandparents would have used, and one of his first business enterprises must have been about 1905 or 1910, when he started a restaurant where the mine Africans in some little mining village on the Witwatersrand would have come to eat.[10]

The immediate economic situation of such immigrants was tenuous. In a 1904-1906 survey of Jewish residents seeking naturalization at the Cape, the largest group by profession was that of traders and merchants, followed by tailors and outfitters, shoemakers, members of the building trade, clerks and shop assistants, traveling salesmen, butchers, cattle-dealers, watchmakers and jewelers, bakers and barbers. There were also a small number of rabbis and scholars.[11]

The flight engendered by pogroms led to the break-up of families in some instances. Wolf Kodesh related the story of his father's family in eastern Europe. At the outbreak of the pogrom the family decided that if it split up, some members might survive:

> So they decided, in their wisdom, that the father, the mother, the youngest girl and their youngest boy would go east, and the three older boys would go west. So that if the Cossacks came, some of them would be alive. I suppose that's the thing that went through their minds. So, the party that went east ended up in Moscow. And those that went west, through Jewish agencies and so on, could have gone to America or South America. They happened to be sent to South Africa [and] landed in a mining town called Benoni.[12]

With the dramatic rise in the Jewish population, organizations were formed to look after the community's interests. The South African Zionist

Federation was formed in Johannesburg in 1898, representing Jews in the four provinces. The Federation negotiated with Lord Milner, High Commissioner at the time, for recognition as the agency acting on behalf of Jewish refugees following the South African War.[13] Jewish Boards of Deputies were formed in 1903 and 1904 representing the Jewish populations of the Cape, Durban, and the Transvaal.[14] The Boards were dominated by English Jews. The Zionist Federation and the Board of Deputies in the Transvaal were antagonists for years. However, Zionism's influence continued to increase among South African Jews and most Board members were Zionists by the 1930s.

If Zionism was considered a cultural movement with political overtones, venues for more overtly political expression existed within the Jewish community. The first Jewish Workers Club (JWC) was formed in Johannesburg in the 1890s by Anglo-Jewish leadership. It closed in 1902. A Jewish Working Men's Club and Night School was formed, also in Johannesburg, in 1896.[15] A Jewish Bundist organization, the "Friends of Russian Freedom Society," was established in Johannesburg in 1905. The divide separating Zionists and socialists centered on the definition of the interests of Jews. Zionists perceived the South African Jewish population as homogeneous and with parallel interests. Jewish socialists saw Jewish society as wracked by the same divisions of class as those visible in the broader society.[16]

A Yiddish-Speaking Branch of the International Socialist League (ISL) was formed in Johannesburg in 1917. The Branch's activities were two-fold: the advocacy of socialism within the Jewish community, and the maintenance of links with the working class as a whole through its major party, the International Socialist League.[17] The Branch was also an important fund raising apparatus of the greater ISL, as in the Branch's acquisition of a printing press for the League.

The radicalization of the Jews who involved themselves in the activities of the ISL in South Africa often had deep roots in the history of eastern Europe. Most Jews there were poor, subject to high mortality rates, and, in the later nineteenth century, fearful for their lives at the hands of the Czar. Thus, the revolutionary politics of socialism were also the politics of anti-Czarism. The General Jewish Workers Union (Bund) was founded in 1897 in Russia, Poland, and Lithuania. Jewish workers came to play prominent roles at both local and national levels in eastern European communist movements.[18]

The history of worker organization and protest was very much alive in the Jewish workingmen's clubs of South Africa. The Jewish Workers Club (JWC), formed in Johannesburg in 1929, represented the apex of socialist or worker consciousness in a particularly Jewish context. The Club served a number of functions "on the Jewish street" and was as much a cultural center as it was a political, self-conscious preserve of Yiddish culture and language. It was also a focal point of social life, featuring activities that ranged from choir get-togethers to a "great proletarian masked ball."[19]

The central anti-Zionist tenet forming the backbone of JWC thinking meant that Jewish workers would, by definition, focus their energies on the state of the working class in South Africa. Only later, from the late 1930s onward, did Zionist socialist organizations synthesize socialist analysis and Zionist initiative. The JWC supported numerous left- and worker-oriented organizations, including the Communist Party, the Friends of the Soviet Union, the Left Book Club, and Ikaka Laba Sebenzi.[20] Members of the JWC also supported the drive for an autonomous Jewish region, Birobidzhan, within the Soviet Union. If there was an apparent contradiction in the dual focus of the South African class struggle and Jewish self-determination in the Soviet Union, JWC members were not troubled by it. The Soviets had overthrown the Czar, an oppressor of Jews. The Soviets were attempting to better the prospects of the Jewish people by creating the Birobidzhan autonomous region, and Birobidzhan flew in the face of the Zionists, who repudiated any home for Jews outside Palestine.[21]

The JWC also had close links with the Communist Party in the 1930s, translating articles of the party's leaders into Yiddish, receiving greetings on its anniversaries from J.B. Marks and the Mofutsanyanas, prominent African communists and union organizers and, during the period of the CPSA's fracturing, guarding CPSA meetings and disrupting meetings of expelled CPSA dissidents.[22]

From the early 1930s the JWC engaged in anti-fascist work. The organization's analysis understood anti-Semitism as a class phenomenon, " 'a deliberate policy adopted by a dying capitalism in its last and desperate struggle to maintain its stranglehold over the working class.' "[23] With the Russo-German Pact of 1939 the JWC switched its line and some members left. Most returned after Hitler's attack on the Soviets and the JWC changed its line. At war's end the JWC's membership declined. There are a number of reasons for this: the shrinking Jewish working class; the movement of younger Jews into non-Jewish leftist organizations; and the ultimate primacy of the Anglo-Jewish culture:

> In the long run it was the Anglo-Jewish pattern which, at any rate in its externals, prevailed in South Africa, although it underwent important changes in spirit and inner content. In other words, the basic trend was for the "Russian" Jews to become acculturated to the older English-speaking section. It was a case of pouring Litvak spirit into the Anglo-Jewish bottles.[24]

This is not to say that eastern European Jewry had no impact on the greater South African Jewish culture. The dominant white cultures of South Africa were Afrikaner and British. Historically, Jews had identified with British cultural and political trends there. This, coupled with the natural process of absorption through language and educational facilities into the dominant culture itself, meant that radical Jewish energies would find other, more generalized venues within which to operate. Younger Jews who sought political roles in the 1940s, as described by Adler, did not want their political activities "deflected and mediated by cultural activities."[25]

Upward Mobility in the Jewish Community

Looking at the occupations of the Jewish informants' grandparents, parents, and the informants' own educations and qualifications, one finds a clear pattern of upward mobility. Where grandparents often started out as managers of eating houses around the mines or as artisans or itinerant salesmen, by the time of the birth of the informants' generation, parents were often the owners of small businesses. A number owned small groceries. Two owned hauling businesses, one owned a furniture shop with his brothers, one taught high school. Two worked as bookkeepers. The parents of three of the informants became quite affluent, one owning a brickworks, another a candy-manufacturing company.

Ben Turok, born in Latvia, related how his parents fled Byelorussia in 1921 to escape persecution. Turok's father had been first a worker and then a shopkeeper there, and with his savings, the family fled to Latvia, where Turok was born in 1927.[26] In 1931 his father went bankrupt, and left for South Africa where the family was reunited in 1934. The elder Turok's political progression mirrors the experience of many Jews who emigrated from eastern Europe and found financial success in South Africa:

> Politically, my father was a Bundist, and a Social Democrat. Really, he didn't like the Bolsheviks very much. He and my mother were loyal and patriotic to Russia, Mother Russia, all their lives, spoke Russian fluently and so on. [But] I grew up in a kind of Yiddish home which I came to reject later. I sort of drew away from the whole Yiddish ghetto life. My father was involved in the anti-fascist struggle in the thirties as a member of the Jewish Workers Club, but after that they really distanced themselves from politics in South Africa, didn't want to get involved. It was a little bit dangerous, and he became a businessman and quite wealthy.[27]

Among the Jews who had arrived earlier in South Africa, some were well established in business prior to the Depression. The reverberations of the Depression were far reaching, and some members of the Jewish community lost substantial businesses in the early 1930s. Among the families that suffered was that of Wolf Kodesh. The Kodesh family had split in eastern Europe, and Wolf Kodesh's grandparents had met in the grandmother's restaurant in Benoni. Kodesh's father had become a cattle dealer and was also involved in transportation with horse-drawn carts, known as "rally carts."[28] Kodesh's parents were not particularly political people, although Kodesh's father took some interest in the Labour Party, and Wolf Kodesh remembered Walter Madeley as a frequent house guest who would bounce him on his knee. During the Depression, his father lost the business and the young Kodesh moved with his mother and siblings to stay with his grandmother in Cape Town. Conditions were fine, but for Kodesh's father "things didn't get right," and his mother was forced to buy a small provisions shop on a street in Woodstock, a predominantly Coloured part of Cape Town: "[This] was a notorious street, almost more notorious than the streets in District Six. You know, you just think of anything that was happening in [this] street, prostitution, drugs, dagga, everything was going on there."[29] The family lived in two rooms at the back of the store, and the environment differed greatly from that experienced by most white South African children. Kodesh's best friend during his teenage years was a young Coloured man, the product of a meeting between a Byelorussian traveler passing through South Africa and a Coloured woman. Among the clientele at the family's little shop were local prostitutes, who took pity on the Kodesh children, and a group of customers, "very nice chaps," who were allowed to come in and go behind the counter to get their own cigarettes and who later were identified as a group of armed bandits. Kodesh had minor brushes with the law himself:

> I used to go and meet friends of mine from District Six, two brothers....We'd go to the houses near to which we lived, which were big houses, and take the dustbins that were left outside, jump over the walls, and spill it all over the gardens -- until the police caught up with us, and drove us around for an hour and then let us go. And we were terrified.[30]

While his parents' marriage never mended, Kodesh's father was able to start a successful new business, which Kodesh puts down to the fact that his father, unlike South African-born whites, did not mind dirtying his hands. The father bought a piece of land and a cart and started making bricks. At the time of his death he left two brickworks and an industrial property in Johannesburg.

Many Jews, like Kodesh senior, were able to overcome adversity and even to prosper. These examples should not be taken to suggest either that by the 1930s there were no working-class Jews left or that a rise in class status precluded continued political activity among radicalized Jewry. Some members of the generation growing up in the 1930s and beyond did so in politicized households, including Denis Goldberg, son of prominent political activists and later a leading member of the Congress of Democrats. Both parents were born in London, although his father's parents had emigrated from Lithuania. A socialist from his teenage years, Goldberg's father joined the British merchant marine to avoid participating in World War I, a war he considered imperialist:

> My father would have said, if he were alive now, as he said to me, the most exciting moment in his life was buying a newspaper in Grand Central Station, in New York, in November 1917, and seeing the triple-decker headlines about the Bolshevik Revolution. That for him was the turning point in his life.[31]

Goldberg's father was an East End London Jew, imbibing his politics there. His ship-board travels took him between England and Australia a number of times before he settled in South Africa in 1929. In South Africa he was a cartage contractor through World War II, with three trucks at one time, and following the war he became a real estate agent. The father was involved in trade union work, and both parents were involved in liberation politics. Goldberg tells a story about his father's real estate agent business in the context of his politics:

> Once he [Goldberg senior] settled in South Africa he was active in trade union work and political work in the national liberation movement. He was a fairly noted public speaker right through World War II. And at the end of World War II he became an estate agent; he was older now, it was a different way of making a living, and he was about to clinch a deal where the United States government would buy a huge home through his agency to establish the home of the Ambassador in Cape Town. The night before the signing of the contract, my dad made a speech attacking the United States publicly; it was reported in the press. Somehow or other, and for some strange and mysterious reason, that deal didn't go through. He got a phone call from the U.S. Consulate to say that they'd changed their minds, they wouldn't be turning up for the signing. That's my background.[32]

Coming from such a background, the problem Goldberg faced was "that of *not* getting involved."

Most parents of Jewish interviewees were not politically active. They were hard-working people focusing on survival and the building of a life in a new country. Radicalism in Jewish families would not necessarily bridge generational changes, as it did in the case above. Amy Thornton remembers her grandfather, a Polish immigrant to London, regaling her with stories of working-class Jewish life in London's East End, where he learned his socialist ideas as a young man: "He used to tell me he remembered Trotsky and George Bernard Shaw speaking in the East End of London, and remembered listening to Shaw in the back of some sort of cart."[33] In South Africa Thornton's father apprenticed in a furniture shop and later opened his own furniture store. Although he introduced her to "ideas of socialism and social justice," neither father nor mother was ever active except for some involvement in a chapter of the Friends of the Soviet Union during the war.

Even though a Jewish child growing up in the 1920s or 1930s might have parents who were concerned to impart ideas of "socialism" or "social justice," the nature of the society in which these immigrants lived, and the force of its dominant ideology, could tend to temper lessons of a more humane nature. Alan Lipman's uncle was a Labour Member of Parliament. The uncle gave him a bar mitzvah gift that opened his eyes to the oppression inherent in South African society: "I was aware of it, certainly, but it didn't strike me as oppression until as a bar mitzvah present I was given Bernard Shaw's *Intelligent Woman's Guide to Socialism.* Which I think was a book that changed my life."[34] Yet the same uncle who gave him a gift that set him on the road to becoming a democrat is described as a racialist whose socialism was mirrored in the fighting slogan of the 1922 miners' strike, "Workers of the World fight and unite for a White South Africa."[35] Of his parents, both of whose families were born in South Africa, Lipman noted that they accepted "the white South African premises and the Jewish white South African premises, which [had] a tinge of 'these black savages are really not quite human.' "

Growing up Jewish in South Africa did not set one on a pre-determined path of humane thought or proletarian activism. It is important to have a clear understanding of the Jewish community's general nature in the twentieth century. By mid-century, given the elevation in class of the majority of the Jewish population, the attendant loss of a working-class identity and disenchantment with the Bolshevik revolution following revelations of anti-Semitism there, the South African Jewish community generally could only be described as conservative. The South African Jewish Board of Deputies was of the opinion that it could not indulge in politics and the making of political statements: There was no Jewish consensus from which to speak. By mid-century the leadership of the Jewish community had defined its role to exclude the making of pronouncements on issues outside the realm of immediate Jewish concern. In 1955 the Board of Deputies did issue a statement from its annual Congress that broached the questions central to any realistic discussion of South Africa and the role of the Jewish community as a responsible component therein:

> Congress repeats its conviction that the welfare of all sections of the population depends on the maintenance of democratic institutions and the enjoyment of freedom and justice by all. It believes that the elimination of inter-group conflict and the abatement of racial prejudice are vital for the national good, and urges support of efforts directed to these ends.[36]

Statements such as the above suggest a kind of nodding gesture of acknowledgment to the black community in its struggle for freedom. At this time the majority of the Jewish voting population was voting for the United Party. To point to the caution of the Board of Deputies is not to detract from what was in 1955, and remains, an on-going debate within the Jewish community surrounding the nature of apartheid, the possibilities for a changed

South Africa, and the community's role in relation to both. What is important to note is the insularity both of the community and the debate. From the post-war period, Jews who pledged themselves to democracy as activists against the white supremacist state did so outside the environment of the Jewish community proper, as individual members of other organizations.

Within the Jewish community, it was generally unacceptable to take positions of clear and direct opposition to the institutions and individuals responsible for the dissemination of racism and exploitation. The case of Rabbi André Ungar is illustrative. Ungar arrived in Port Elizabeth in 1955 at the age of 25 to head a Reform congregation. Ungar spoke emphatically against apartheid and its various manifestations, such as the Group Areas Act, both within the synagogue and in broader public forums. In 1956 Ungar's residence permit was withdrawn. While some spoke up to defend Ungar, others castigated him. A Jewish letter writer to the *Port Elizabeth Evening Post* on 21 December 1956 called him "a non-desirable visitor....[who] lacks the sense of responsibility and dignity of a responsible leader of a community."[37] The Eastern Province Committee of the Jewish Board of Deputies specifically denied the suggestion that the government's action was an attempt to intimidate Jews critical of its policies. The national executive of the Board stated that Ungar "went on to the political platform and must therefore bear the consequences as an individual," so reinforcing the Board's policy of non-intervention in matters political.[38]

It is clearly insufficient to point to historical precedent or Jewish cultural traits as explanatory vehicles for the prevalence of Jewish dissenters in South Africa. It is insufficient because such dissent, as it applied to agitation for a unitary, democratic state with respect and equal rights for all South Africans, was not general within the Jewish community. The Jewish community was undoubtedly perceived, particularly by Afrikaners, as peculiarly liberal. Yet its insularity, the fruit of its fears and the manifestation of its desire to live and develop unhindered in South Africa, guaranteed that members who chose to enter the political arena and speak in terms outside the boundaries of acceptable dissent would not be welcome as members of the community. One must then ask: What led a minuscule minority of the Jewish community to forsake the privilege that was its natural right as "South African white" for identification with the struggle of the Indian, Coloured, and particularly African members of that society?

The influences upon the sample below were numerous and complex, and an attempt to categorize and quantify them in a number of neat psychological or experiential categories risks oversimplifying both individual and collective histories. These are experiences of awakening, often powerful and immediate, and with serious ramifications. For nearly all of those people interviewed, such experiences led ultimately to banning or detention or long-term imprisonment and exile. Some of the themes that arise throughout the interviews include the influence, during childhood, of black people who were in the young person's life in an abiding sense, although this is not a specific designation for "nannies." In a more general sense still, a number of those interviewed grew up in communities for at least a part of their young lives where black and white lived together and played together, as in the case of Kodesh in Cape Town. An instance was cited above where a book led to a moment of insight about the nature of South African society and the oppression inherent there. If there is a single particularly striking motif throughout these interviews, it is the power of the written word and its influence on this group. A number of "awakenings" are attributed to something read, which suggests that this generation, unlike its predecessors, did not have the direct experience of the proletarian battlefield. It is not true to say that this generation had *no* organized political experience prior to its making a full-blown commitment to political activity, only that such experience was more likely to have been in university clubs, youth organizations or political bodies, rather than in unions.

Hymie Barsel's mother instilled in him a keen sense of right, justice, and morality. Growing up in an Orthodox Jewish family, he attended Synagogue in Fordsburg every morning, and religious holidays were observed at the Orthodox Synagogue in Doornfontein. Childhood was spent working in the family shop or playing football out in the street with the children of the neighborhood, black and white. Following an accident that left him epileptic, Barsel was brought by his sister to the office of Dr. Max Joffe.[39] Joffe treated Barsel, and one day told him about the Youth League of the Communist Party, of which Joffe was chairman. Much of the Youth League's focus at the time was on anti-fascist work, and Barsel helped around the office and attended the functions of the group.

A demonstration by the Blackshirts, an organization of Nazi sympathizers, provided an important moment in Barsel's political development. Joffe called him and told him of a planned demonstration by the Blackshirts, the object of which was to smash Jewish shop windows. Joffe told him to come and fight, which he did, and there he was surprised to "[find] myself next to the Greek, Portuguese, German, all non-Jews. I had expected only Jews to interest themselves with anti-Semitism fights." During the demonstration Barsel was struck on the head by a policeman and went to Joffe's office for first aid. On the way he was joined by a young Coloured man who had also been struck. Barsel asked him if he had been caught in the crossfire, and the young man replied that he had come to fight "the bloody Nazis." The young Coloured man told Barsel that anti-Semitism was just one form of racialism, and one must fight all forms of racialism, not just that aimed at Coloured people, to realize freedom:

> You must destroy all forms of racialism, must destroy all forms of racialism. Anti-Semitism won't be destroyed unless you destroy all forms of racialism. I could think of nothing else all day. I went to Max's consulting rooms. I

> discussed it with him and yes, he said, Nick is a communist, a good chap. It is right. But he suggested that I read some books.[40]

Barsel joined the Left Book Club, a reading and discussion club, and read a great deal about the anti-fascist struggle and the Soviet Union. He did not support South African entry into the war until the Nazi attack on the Soviet Union, when he attempted to enlist in the army and was rejected for health reasons. In the meantime he was offered the job of founding *Die Ware Republikan*, a Communist Party paper specifically aimed at the Afrikaans-speaking working class.[41] The paper failed in 1943, and Barsel became a rent collector for the City Council in Western Native Township. It was customary practice among the white collectors to wait to write out rent receipts until six or eight Africans were in the office, an inconvenience for Africans but a means to concentrate the clerk's time for other tasks, or for leisure. Barsel thought this unfair and wrote out receipts on the spot. He was fired in a matter of months:

> Jessie[42] came back to me, she said she raised it [Barsel's dismissal] in the city council last night; they said I was a communist and stirring up trouble among the natives, Africans as they were called then. Ja, being civil to black people made me eligible for dismissal in my job, civil to black people made one a communist. This was 1942 and many blacks were in the armed forces fighting for South Africa.[43]

After his dismissal, Barsel went to work full-time for the Friends of the Soviet Union. Perhaps Barsel's entry into political activity, beginning as it did with the Communist Party's Youth League, seems like a precipitous leap. Still, it is important to remember that many members of this generation, including Barsel, were baptized in the anti-fascist struggle, which for young South African Jews was an immediately compelling cause. Only following this initial immersion through the experiences of the anti-fascist struggle such as the one noted above, did Barsel's political vision grow to encompass a broader concept of struggle based on an analysis of anti-Semitism as a form of racialism, which was itself tied to class struggle and imperialist design. These realizations were, at least in part, the result of study.

Books and personal experience were important in the radicalization and democratization of the interviewees. Witness Rusty Bernstein. Bernstein was born in 1920 in Natal, the child of an eastern European father and French mother. Orphaned at a young age, he knew little about his parents' early histories, save that his father had been a bookkeeper in a factory. To the best of his knowledge, neither parent had been politically active. He explained his political radicalization as the result of years spent in a public-type boarding school, Hilltown College near Pietermaritzburg, in the mid-1930s. The school imported its teachers from the ranks of young men at Oxford and Cambridge, which Bernstein noted at the time as having produced a "rather radical anti-fascist generation." His first organized political experience took place as a member of the Labour League of Youth (LLY), the youth wing of the Labour Party. There he met his future wife, Hilda Watts. Both had also joined the Communist Party, and both had to choose, along with all CPSA members who were also members of the Labour Party's Labour League of Youth, between the two organizations. Dual membership would no longer be allowed. For Watts this was not a difficult decision:

> We thought in the Youth League of the Labour Party that we would be able to influence the LP to a more acceptable attitude towards the black people....and we just couldn't do it. There was a city council election when I was a member of the LLY, and a butcher in Johannesburg named Light stood for the LP under the slogan, "vote for Light and keep your city white." So I think it was our general feeling of frustration and inability to get anywhere...It was not a hardship for me to leave the LP.

A number of informants experienced their first organized political activity in Zionist-Socialist groups. Zionist-Socialists diverged greatly from the Zionists proper on the matter of political activity within South Africa. The Zionist-Socialists declared their alignment " 'with the socialist labour movements of all countries,' " which led them to participate in activities with many groups on the left.[44] Ellen Hellman was drawn to the Zionist-Socialist program, and explained its pull for the young Jewish generation to an American audience in 1944:

> The enormous inequalities between the different racial groups in our race-caste society and within the white group itself, and the very real and threatening menace of the opposition party have had the effect of arousing in young people a turmoil and ferment and a spirit of revolt which seeks release in leftist activity. A number turned to the Communist Party and others sought an outlet for their drive towards change in other organizations. General Zionism with its emphasis on emotion could not hold them. Consequently Zionist-Socialism with its clear-cut ideology, its synthesis of Zionism and Socialism, its affirmation of both the *Yishuv* and the culturally potent *Galut*, has been able to provide a Jewish political home to many who were restless and frustrated or entirely remote and aloof from the Jewish scene.[45]

Yet the Zionist content of their organization led them to hold that a national homeland for the Jews in Palestine was necessary for the national existence to be "normalized" and therefore was a prerequisite to the realization of Jewish socialism.[46] Harold Wolpe's parents, Lithuanians, arrived in South Africa before the Boer War. Neither parent was particularly political, but both were hostile to British imperialism as it impinged on the Palestine question. Wolpe, along with a number of others interviewed, attended Athlone High School in Johannesburg, which was situated in an area that was home to much of

Johannesburg's Jewish working class. The Jewish Workers Club originated there, and Wolpe, who later joined COD, the South African Communist Party, and Umkhonto we Sizwe, remembered the weekend sidewalk activity with its non-stop political discussions. Entering university in 1944, Wolpe looked for an organization in which to become involved. He taught literature classes in a Zionist-Socialist night school for Africans. To get to the classes he would take the tram across Johannesburg two or three nights a week, and when he came across *What Is Marxism* by Emile Burns, he brought it to read on the tram. He said it provided "what I'd been looking for." At the end of his first year at University he joined the Young Communist League and began to attend meetings in the townships.

Michael Picardie's story typifies the diverse elements in the making of Jewish radicals. He credits his early sensitization to the fact that, as a child, he spent much time with and looked up to a group of African men. Picardie's mother built blocks of flats, which the family would live in until all the units were sold. She employed African migrant workers from Natal to do much of the construction work, and Picardie identified with these African men:

> Black men came to seem to me to be an important source of my identity. They were all of one family, Ngubane, from a village called Inyanga Stores, near Greytown, and generation after generation came. There was the old man, who was the watchman, and then there were three or four who were domestic workers, cleaning the flats, and one or two young men, just a few years older than me, and as I grew older a few my age, so I saw people ranging from about 50 or 60 to about 17 or 18. All from the same family or clan, and I began to feel at home inside African society.

The British-style public school that Picardie attended, King Edward's School in Johannesburg, had a radicalizing effect on Picardie as it had had on Bernstein. The student body was composed of the sons of men involved in Johannesburg's professional and commercial world, hailing from suburbs such as Belleville and Yeoville. Still, to Picardie it was a cosmopolitan world where Zionism and politics were discussed outside the classroom and a small number of teachers encouraged the students to think critically:

> This one history teacher, Teddy Gordon, taught us the French Revolution in such depth that we were able to begin to understand the dynamics of class conflict and the nature of the state....I began to realize that the understanding of the French Revolution that he was helping us to acquire could be applied to understanding why the Nats were destroying the Coloured vote in the Cape. We were beginning to be aware of what the Boers were doing so it just needed one more step to say, not only are they racists vis á vis the Jews, but the Jews themselves are racists vis á vis the blacks. And we must detach ourselves from our Jewish identity and start identifying with the democratic identity that the boys had brought back from up north, the Springbok Legion.[47]

Picardie understood why most Jews did not make that leap: They carried fears with them from other countries, fears that grew to a much greater pitch after the Holocaust.

Picardie finished at King Edward's in 1953 and entered the University of the Witwatersrand in January 1954. Here, Picardie and his friends from similar backgrounds met blacks with equal educational experience for the first time, blacks who were involved in the student politics of the university. Picardie's politicization was in part a result of his relationships with these blacks, many of whom who were involved at that time in the Western Areas Campaign. He was elected a member of the Students' Liberal Association (SLA) committee, and many of the students working in the SLA were also involved in the Congresses or had been members of the Communist Party. These students introduced him to the theoretical discussions surrounding the liberation struggle:

> Joan [Anderson, a member of COD] introduced me to political thought via two books, Harold Lasky's account of Marxism in a small book, I think it was a book called *Communism* by Harold Lasky, and the other book was I think Isaiah Berlin's biography of Marx. Lasky's book turned me virtually overnight, or within a few days -- I had a kind of conversion experience.[48]

For those who came from radical backgrounds there were, of course, no conversion experiences. Denis Goldberg came from a family where involvement in political activity was taken for granted. Goldberg grew up in Cape Town and was the first in his family to attend university. The police would raid his house looking for evidence of his parents' political activity, but this was not his world. He entered a degree course in Civil Engineering at the University of Cape Town. From boyhood, he had marveled at the great products of civil engineers, the bridges, the aqueducts, the dams. His parents' friends asked him why he was not involved politically, and other students of similar background would jokingly suggest to one another that someone should tell Goldberg how to vote in the Student Representative Council elections, as there was a chance he might vote correctly. It was in 1953, while working on his undergraduate thesis, that Goldberg knew he was going to have to get involved:

> Because by then, I'd come to understand that you couldn't be a civil engineer in apartheid society. Everything you did was political. If you lived in Cape Town you had to know that the Cape Town railroad station was costing millions of pounds more because it had to be an apartheid station. The platforms had to be placed illogically from an engineering point of view because black lines had to go to black townships, and white lines had to go to white suburbs, and I couldn't but be aware of it.[49]

Goldberg graduated in 1955 and took a job with the railroad. He was fired shortly afterward because he was working on the organization of the Congress of the People at the same time. He would lose another job in 1960 for similar reasons. His political activities culminated in imprisonment.

Norman Levy, as with Goldberg, grew up with a politicized parent. Levy's family lived in a working-class Jewish neighborhood in Johannesburg, and his father was involved in socialist activities, but Levy was not aware of this as a child. His radicalization took a separate path. It was a reaction to the racism and economic inequality he found in Johannesburg rather than a reaction to his own class position or that of east European Jews as an immigrant group. Levy entered the Young Communist League as a teenager, which, like the Communist Party, was a non-racial organization, and there Levy had his first extended contact both with blacks and politics.[50]

For many of these Jewish interviewees, entry into an organization such as the Young Communist League or the Left Club came as a revelation. Amy Thornton and Ester Barsel both entered Hashomer Hatz'air, an organization for young Jewish leftists who expected ultimately to immigrate to Israel. Their experiences in Hashomer were different, but neither found a political home there. Thornton said she "fit in best" in Hashomer of all the Zionist organizations she had entered, given its leftist bent. It was only when a high school friend invited her to join the Left Club she realized that "there was a struggle going on in this country." There she met blacks for the first time "on an equal level." She attended political classes and shortly thereafter joined the Young Communist League. Barsel's experience in Hashomer was somewhat more radicalizing. She participated in Marxist study classes there, and the prevailing ethic was one of disdain for "bourgeois things." She was not interested in settling in Israel though, and after a trip to England, she became involved with the group of progressive whites in Cape Town.[51] This group formed a loose weekly lecture collective, which became the Modern Youth Society.

The South African Jewish community was therefore, by the mid-twentieth century, very much a hybrid. The majority of its members were of eastern European origin, but those eastern European Jews were relative newcomers. Though in the numerical minority, the Jews who had arrived from England, Germany, and other nations of western Europe were much more readily assimilable; even the Jews of the Pale found themselves assimilating over time. But assimilating to which culture? To the liberal British imperialist legacy? To an identification with the culturo-religious nationalism of Afrikanerdom? Or to a generic sort of "white culture," in which prior cultural identification was obscured in the process of realizing the power and dominance one enjoyed on the basis of one's skin color? No simple answer presents itself, and one can point to examples that would fulfill any of the above categorizations. For the majority of South African Jews, none of the foregoing identifications would be wholly meaningful. Yiddish culture was not forgotten but was marginalized, as Jews broke out of the working-class ghetto. Anglicization was a definite part of the Jewish immigrant's reality: He became conversant in English and so tended to identify with the English political standpoint. Yet, the Jew was not an English. And in a country where anti-Semitism was acceptable to the majority of the white population, and while the Holocaust was a living reality, he could not forget his heritage. This is not to say Jews hated Afrikaners. They felt the potential threat of the Afrikaner's anti-Semitism, yet there was room for grudging respect. Afrikaners could both respect and empathize with the Zionist dream, and Jews did find allies in the Afrikaner community. Afrikaners saw in Jews "descendants of the Biblical children of Israel," which Shimoni suggests fostered a certain degree of empathy. The National Party wooed Jewish support in the 1924 election. Tielman Roos, the NP Transvaal Party chair claimed that Afrikaner nationalism should appeal strongly to Jews, given the "tenacity of purpose" exhibited by Jews, and by inference, Afrikaners. Roos defended state provision of facilities for raising Jewish children in the Jewish faith and culture. Shimoni, noting the historian A. Keppel-Jones's prediction in 1947 of an anti-Jewish pogrom by the mid-1950s, described the Jewish-Afrikaner relationship as one of gradual "accommodation and rapprochement," tested, really, only by Afrikaner perceptions of a hostile Jewish minority working to defeat Afrikaner hegemony:

> When photographs appear in newspapers of resistance processions, or of joint singing and dancing with the "Africans," or of the Black Sash's slander tableaux, the Jewish facial type is in the majority. When a book is published on the "bad conditions" in South Africa, the writer is ten to one a Jew. Under petitions protesting against the Boer's policy there always appear numbers of Jewish names. Jewish professors, lecturers, doctors, rabbis and lawyers fall over one another in order to be able to sign. Behind tables in the street collecting signatures against the Boer's policy a Jewish lady is usually enthroned....The patience of the Boer must one day become exhausted.[52]

Despite these perceptions, the Jews kept largely to themselves. The Jewish community was an insular community. If the Board of Deputies represented the community generally, its ethos was one of protection, or normalization of relations between that community and the wider white community. It sought to protect the interests of South African Jews, and it did so primarily through a policy, through the 1950s, of non-involvement in the issues facing the greater South African society.

How then does one explain the commitment of the individuals profiled here? A number of possibilities suggest themselves. Jewish culture may have played a part, as indeed a number of interviewees themselves suggest.[53] The stress on education, on self-reflection at an early age, and a consciousness of a history of suffering at the whim of others acted to sensitize these people. Also, these grandchildren and children of the victims of pogroms grew into conscious young adults during the fascist era.

For the Jews of Lithuania and Byelorussia, and for the Jews of London's East End, there existed a history of worker struggle, a legacy of organization and education in socialist thought to draw upon. Jews were among the founders of radical and worker-oriented organizations in South Africa. And if these children of the working class were able to enjoy a better education and a more affluent future than had the generations before them, they remained aware of the rot in the society around them. The rise of fascism and the fight against it, in which Jews played their part, further exposed the hypocrisies of South African society, as the wartime slogan, "the fight for democracy for all," was measured against the post-war reality for blacks of conditions unchanged and promises unfulfilled.

These social, cultural, and economic factors are broadly explanatory of the motivations for Jewish involvement in the national liberation movement in the 1940s, 1950s, and 1960s. Individual motivations are more difficult to grasp: in some instances a book read, in others the joining of a club, the flaring up of a sense of injustice and indignation, the fight to protect one's own culture, or one's life.

One might be moved to ask why more Jews did not become involved in anti-apartheid activities. Fear was certainly a motivating factor but probably not the most important one. Perhaps the primary factor was prejudice. The majority of Jews in South Africa were more liberal than their other white counterparts, given their culturo-historical experience, but they also shared many of their prejudices. Progressive, democratic Jews, like democratic Protestants or Catholics or Afrikaners, were in very short supply. This disparity of numbers points up how exceptional were the individuals we have discussed here.

Afrikaners and Other Heretics

The number of Afrikaners who chose to identify with the struggle for liberation and democracy through the 1960s was small relative to Jewish participation. Immigrants from the United Kingdom played a major role in the rise of socialist thought, working-class politics, and labor organization in South Africa. This section will examine in some detail the backgrounds and early lives of some of the actors in each of these groups.

Bram Fischer

Although few Afrikaners could be counted among the ranks of "democrats," a substantial number of those who could be so counted stood out in their areas of endeavor. Perhaps the most prominent Afrikaner radical of all was Bram Fischer, born in the Orange Free State in 1908. Fischer's grandfather had been Prime Minister of the Orange River Colony, his father a Judge-

President of the Orange Free State Supreme Court. Following family tradition, he attended Grey public school and Grey College, and took a law degree from the University of South Africa, of which Grey College was a constituent part.[54] Mary Benson, herself a South African-born journalist and democratic sympathizer, came to know Fischer well during the Treason Trial, the trial of 156 Congress Alliance members in 1956, and described him as "*n ware Afrikaner*," a true Afrikaner, a lover of the veld.[55] It was Bram's grandfather, as Minister of Lands, who was primarily responsible for the enactment of the 1913 Natives Land Act, the primary legislative act responsible for the dispossession of the African people.[56] At Grey, Fischer refused to wear the school uniform because it was British in style. On the occasion of his twenty-first birthday in 1929, Fischer received a letter from Mrs. Steyn, wife of the President of the Orange Free State, which said of Bram:

> *As kind en as student was jy 'n voorbeeld vir almal, en ek weet dat jy not 'n eervollig rol gaan speel in die geskiedenis van Suid Afrika* (As a child and as a student you were an example to everyone, and I know you will play an honourable role in the history of South Africa).[57]

Mitchison recounts how Fischer, as a young man, attended a meeting of the Joint Council of Europeans and Africans, where he took a black man's hand and felt a strange revulsion that caused him to question "the very bottom of his own feeling."[58] In 1931, Fischer was awarded a Rhodes scholarship and spent the next three years at Oxford. He also traveled as a student representative to the Soviet Union, Bavaria, and Switzerland. These years and experiences apparently were a key formative political and philosophical time for him. He witnessed the rise of fascism and the increasing popularity of British Labour politics. It is impossible to pinpoint the moment at which Bram Fischer could envision for the first time a non-racial, socialist society. In 1943 he helped the African National Congress draft its new constitution, and in the 1940s he was on both the Johannesburg District Committee and the Executive Committee of the CPSA.[59]

It is possible to piece together a sense of the forces driving Fischer to such a rejection of his heritage. In his letter to the court at the time of his journey underground, Fischer stated that he was not motivated by fear of punishment but by a sense of responsibility to bring people to think about the apartheid policies that were being propagated: "To try to avoid this becomes a supreme duty, particularly for an Afrikaner."[60] He found sadness in the fact that the actions of the Afrikaner people were turning Africans against Afrikaans. He saw in the Afrikaner past the potential for progressivism, for support of the working class. He told a story of having lunch with Verwoerd in the 1940s, when the latter was attempting to decide whether to be a Nationalist or a Socialist.[61] Benson suggests that, in a sense, Fischer was a one-man act of penitence for the wrongs of Afrikanerdom. His actions were inspired by his heritage.

Prior to sentencing in 1965 Fischer made the customary statement from the dock. He spoke of time he had spent in the African location in Bloemfontein where he taught reading and writing. There he "came to understand that colour prejudice was a wholly irrational phenomenon and that true human friendship could extend across the colour bar...That I think was Lesson No. 1 on my way to the Communist Party."[62] He spoke of how the Afrikaner had cut himself off from all human contact with the black South African majority and of how the preceding fifteen years of apartheid had produced "evils and humiliation" that could only be laid at the Afrikaner's door: "All this bodes ill for our future. It has bred a deep-rooted hatred for Afrikaners, for our language, our political and racial outlook amongst all non-whites -- yes, even amongst those who seek positions of authority by pretending to support apartheid. It is rapidly destroying amongst non-whites all belief in future co-operation with Afrikaners."[63]

Bram Fischer attempted to serve the national liberation movement in two particular capacities: as an activist and as an advocate, defending political prisoners, including himself, in South African courts. Fischer was a member of the Communist Party and had led the defense in the Treason Trial, the trial of 156 Congress Alliance activists following the Congress of the People and ratification of the Freedom Charter.

In 1964 Fischer was arrested for his political activities. He demanded and was granted the right to leave South Africa for the purpose of finishing a case before the bar in London. Fischer promised to return, and he did. However, shortly after the start of his own trial, Fischer disappeared, stating in a note to his counsel, "I owe it to the political prisoners, to the banished, to the silenced, and to those under house arrest, not to remain a spectator, but to act."[64] Months later, Bram Fischer was captured by the police and returned to trial, charged under the Sabotage Act. On 5 May 1966 he was sentenced to life imprisonment. Fischer was diagnosed with cancer in 1974, and was only allowed to leave prison a short while before his death, to stay with his brother in the Orange Free State.

Bram Fischer must have been a frustrated man. He could find in his own people's history the kernels of humanism and progressivism that they themselves did not see and could not begin to extend to others. Bram Fischer was a man very much respected by those around him. He was a leader, but he was an open man, and the house that he and his wife Molly shared was the movement's house, a kind of non-racial swimming oasis.[65] Until his journey underground, arrest, and life imprisonment, Bram Fischer lived his life outwardly as a respected member of the elite South African community. Yet his life stood as a direct challenge to the will of "his people," and in the end he was disowned by them.

Piet Beyleveld

Piet Beyleveld was probably the most active Afrikaner in the organizational realm of the national liberation movement. Beyleveld was born, like Fischer, in the Orange Free State, in 1916. He attended farm schools and farmed in Namibia between the years of 1935 and 1938.[66] There is little documentary information about Beyleveld, as with most of the whites who figure in this study. There are no writings on his early life and the reasons for his movement into the democratic camp. When South Africa entered World War II, Beyleveld was one of the founders of the Springbok Legion. Beyleveld held myriad leadership positions in workers' and leftist organizations. He was the general secretary of the Textile Workers' Union and was appointed vice-chairman of the Textile Workers Industrial Council. He was a national organizer for the Labour Party. In 1953 he became the first president of the South African Congress of Democrats. He was also the first president of the South African Congress of Trade Unions (SACTU), in a sense the satellite union of the Congress Alliance, upon its formation in 1955. In testimony given later he acknowledged becoming a member of the South African Communist Party, then underground, in 1956.[67] He was also a leading organizer of, and participant in, the Congress of the People. Piet Beyleveld was, one might say, the token prominent Afrikaner in the legal national liberation and union organizations of the time. Arrested in 1964 under the 90-day law, Beyleveld became a witness for the state against thirteen of his colleagues. One of those colleagues, and a friend, was Bram Fischer. Beyleveld, while giving testimony that would ultimately play a part in Bram Fischer's receiving a life sentence, acknowledged his reverence for the man.[68] That Beyleveld betrayed his peers was almost unbelievable to those who had worked with him.[69]

Bettie du Toit and the Sachs "School"

An important point of contact between Afrikaners and radicals both black and white was the union movement. Specifically, one can point to the E.S. "Solly" Sachs "school" of Afrikaner women unionists trained during Sachs's tenure as head of the Garment Workers' Union, from 1928 to 1952.[70] Sachs was a Lithuanian Jew who arrived in South Africa in 1914, joined the ISL in 1917, and became involved in union activity in 1920, at which time he opposed the organizing of white and black in one union. This position put him at odds with other members of the Young Communist League, including Eddie Roux.[71] Though a member of the CP for some time, Sachs stubbornly resisted programs involving non-racial organization and chose to defend his organizing of parallel black and white unions as a more successful vehicle for wage negotiation.[72] Sachs was not a racist. He worked with leading Africans of his time, including Gana Makabeni and Lillian Ngoyi, perhaps the foremost woman activist of the 1950s. Sachs, Weinbren, Kalk, and other white union organizers have been credited with neutralizing what might have been the expected pro-fascist and racialist tendencies of their Afrikaner workers as a result of their strong unionism and socialism.[73]

Bettie du Toit described Sachs as a "brilliant" tactician. Du Toit was a second generation product of the Sachs "school," brought into union work by Johanna Cornelius, sister of another Sachs disciple, Hester Cornelius.[74] Du Toit, an Afrikaner recently arrived from the countryside, was working in the Johannesburg area at the time. She attended a meeting of the Friends of the Soviet Union at the Trades Hall, where five young Afrikaner women who had recently visited the Soviet Union were reporting their experiences.[75] There, Johanna Cornelius approached her and asked if she would help with the strike then taking place among textile workers at Industria. Du Toit agreed and was subsequently arrested and driven away with five other young Afrikaner women in a police van. Du Toit and Cornelius refused to pay the fine levied against them for their protest activities and were jailed. Shortly thereafter du Toit was asked to travel to the Cape and organize textile workers there in the 1939-1940 period.

Du Toit's first assignment was short-lived. The textile workers' committee was led by a white secretary and a Coloured chairman. Dances were often held as fund-raisers, and at one of these dances:

> one of the men asked me to dance with him, and unthinkingly I danced with him. He foolishly went into the factory the next day and said, "oh, Bettie du Toit is not like you, she danced with us last night," so this created a stir, and I lost a lot of white members.[76]

She then moved on to Paarl in the Western Cape where she worked in a textile factory and lived in a National Party hostel with other young Afrikaner women. Interestingly, the job was procured for her by a prior foe, a Jewish textile owner by the name of Moburger who had defeated du Toit and the union at Industria but who, perturbed at the development of fascism in South Africa, saw reason to have du Toit working among such women. The work went well until word spread about her activities and reached the Dutch Reformed Church. When representatives of the Church searched her personal effects, they found a book dealing with the Colonial and National Question. At some point between the late 1930s and the early 1940s du Toit had joined the Young Communist League. Returning from a short trip to a Communist Youth League function, du Toit found that no one would speak with her. The newspapers spoke of "Communism in Paarl." Conditions became untenable and du Toit left for Cape Town. There the Minister of Labour called her in to ask what she planned to do, as a public meeting had been called by the Greyshirts (Nazi sympathizers like the Blackshirts) to discuss how to meet the communist menace. He suggested she attend, which she did, with her fellow unionist and friend Joey Fourie:

> They were ranting and raving and explaining how wicked and terrible we were....Sitting behind us, one of the farmers said in Afrikaans, "but where is this terrible Communist?," so I turned around and said very meekly, "it's me sir." So he said in Afrikaans, "Man, but you're very pretty!" and the whole hall shrieked with laughter....I got up and spoke at the meeting and nobody touched us.

By the early 1940s Bettie du Toit was radicalized, an Afrikaner communist who unthinkingly danced with black men. When asked how she had come to lose her racial prejudice, which she acknowledged as having been of the type and magnitude one found on the platteland, du Toit could say only that it was a function of time and circumstance, time spent with workers and their problems, visiting them in their homes. Bettie du Toit recounted one moment of realization:

> I was at a South African Trades and Labour Council Conference, and I wanted to attract the attention of H.A. Naidoo because I had been active with him in Durban, organizing the Indian workers there. Between us, sitting, was an African male, and I leant across him and pulled H.A.'s sleeve, and I told him whatever I had to tell him. Then a short while after that a mineworker got up, went to the rostrum, and attacked me and said, not only does she talk to these black men, but she leans across the one and pulls the sleeve of the other! I thought, I really have got rid of my racial prejudice.[77]

While Bettie du Toit was never a member of the Congress of Democrats, she linked worker organization and economic issues to broader questions of workers' relationship to the political realm and to the apartheid structure for those she organized. She recollected putting a resolution against passes before the members of an African branch she was working with. "[They] said to me, [referring to passbooks], who's going to identify me, what am I going to be identified with?" The resolution was rejected. Bettie du Toit was one of the handful of whites who participated as resisters in the Defiance Campaign, and after serving a short jail term she was banned for life from further trade union work.

In the mid-1950s du Toit opened the People's Cooperative Trading Society in Orlando East, and toward the end of the decade she became involved in a school feeding scheme for African children in the townships, known as Kupugani. Du Toit was welcomed in the townships, and the scheme fed children in 26 schools at one time. Working with a permit under a false name, she was finally caught at work in the township and left the country in 1964, which saddens her still. Bettie du Toit said of Afrikaners that they were, to a certain extent, vicious, but if convinced they were right and on the side of right, nothing would move them. Bettie du Toit, Bram Fischer, and Piet Beyleveld were exceptional individuals. By moving nearer to an identification with the vision of a racially and culturally unified South Africa, they moved so far from the conventions of Afrikaner nationalism at that time as to become culturally unrecognizable.

British Immigrants

Many radical British immigrants to South Africa were labor organizers. From David Ivon Jones and S.P. Bunting in the first decades of the twentieth century, British intellectuals and organizers brought with them a wealth of theoretical knowledge and experience, in the Marxist canon and on the picket line. A number of these individuals would play key roles in the national liberation movement of the 1940s, 1950s, and 1960s. Yet it would be misleading to characterize this group solely as members of the working class. Among the whites making the greatest contributions in this period were clerics and others who came to engagement in the struggle against apartheid by diverse paths, paths that differed greatly from those of their working-class colleagues. Fred Carneson is an example of the path taken by the British working class into liberation politics. His father was a railway worker, a trade unionist, and a Labour Party supporter. While working as a messenger Carneson was introduced to Marxist literature, started a study group with friends, and joined the Communist Party. Percy Jack Hodgson, like Carneson, was born in South Africa, and his father was killed in a mine accident. His stepfather was also a miner. When Hodgson was thirteen he left school and began working around the mines, where he met a number of socialists. Hodgson traveled to Northern Rhodesia and played an important role there in the unionization of African mineworkers. Hodgson joined the Communist Party, was a leading member of the Springbok Legion, and a founding member of the Congress of Democrats. The experiences of Carneson and Hodgson are not unique. Both were grounded in working-class experience, and aware of the importance of organized labor, and each was an avid reader capable of imbibing ideas from the works of Jack London or the publications of the Communist Party.

This section will not, however, focus on the particular experience common to the British working-class generation in this period. Rather, to appreciate the diversity of background and experience found among those of British background who identified with, and became identified with, the national liberation movement, two of these individuals, Helen Joseph and Father Trevor Huddleston will be profiled. Both are remarkable, not just because it may appear surprising that they found their way into the movement of the 1950s, and became prominent there, but because both sought ways to maintain that activity, within South Africa or without, into the 1990s, and so stood as important symbols of the possibility of tenacity and commitment among white South Africans toward the ideals embodied in the national liberation movement.

Helen Joseph

Helen Joseph was born into a middle-class family in Sussex in 1905. Though a confirmed Anglican, she was educated at a Catholic convent school, a choice Joseph described as probably an effect of her family's "middle-class snobbishness." She ultimately took a second-class honors degree in English at the University of London. At the age of twenty-two Joseph left England for a teaching position at a school for Indian girls in Hyderabad.[78] Joseph loved the school and her students but admitted freely that she arrived ignorant of Indian history, culture, or politics, and left largely the same way. The building agitation for independence on the part of the Indian people did not enter Joseph's consciousness in any particularly meaningful sense, albeit Gandhi and Nehru were prominent figures by that time. Hundreds of thousands of Indian workers were taking part in strike actions during the period of her stay in the late 1920s. When Joseph was injured in a horse riding accident, she decided not to extend her contract with the school. Unable to find other suitable work in India, she decided to go to South Africa, where she had a friend in Durban whose father headed a small preparatory school for boys. She took work at the school and shortly thereafter met and married Billie Joseph, Jewish and seventeen years her senior. Helen Joseph's life at the time offered no intimation of changes to come: "We married and moved into Durban society. I rode, played tennis and learned to play bridge. We lived in an attractive Spanish house in Durban North, high on a slope looking down to the sea and across to Durban and the Bluff headland five miles away. I loved my garden, spending much time in it."[79]

With the outbreak of war, Billie Joseph, a dentist, left to join the Dental Corps. Helen Joseph joined the Women's Auxiliary Air Force (WAAF), having seen a newspaper advertisement recruiting university and professional women to a training course for welfare and information officers, to become female concomitants of male Army Education Service (AES) workers. No one was guaranteed a post upon completion of the course, but Joseph was selected, the mission of the officers being to instill in the South African women recruits " 'a liberal, tolerant attitude of mind' " consonant with the anti-fascist goals of the war itself, which Joseph described as truly "an astounding mission."[80] Posted to Bloemfontein in the heart of the Orange Free State, Joseph was charged with lecturing on a wide range of topics that included liberalism, socialism, communism, and "Native," Indian, and Coloured affairs. None of these subjects had been of particular interest or concern to Joseph prior to taking up this employment, and she decided it would be necessary to educate herself before attempting to educate others:

> As I studied the conditions in which black children struggled for education and opportunity, and compared that to how most whites lived, I began to feel ashamed of my own position as a white. Talking about democracy brought home to me that the black people did not share it with me. I had a parliamentary vote and they did not. As I spelt it out to the WAAF's, so I spelt it out to myself, questioning my own values as never before. I did not turn immediately into a socialist, far from it, but I began to see people as human beings, regardless of colour, began to have some idea of how the other half of the South African world lived.[81]

Returning from her wartime assignment, Joseph ended her marriage; actually, both acknowledged it. She then entered a course in sociology and social work at the University of the Witwatersrand while working in a Johannesburg community health center for poor whites. Joseph enjoyed that position, but, as evidence of her increasing consciousness of the deprivations of blacks, she applied for and was hired as a community center supervisor among the Coloured population of the Western Cape. Undoubtedly this experience played a part in the gradual transformation of Helen Joseph. She found love and acceptance in the Elsie's River community, and her political insights deepened, setting the stage for her next position in 1951, working with Solly Sachs as Secretary Director of the Medical Aid Society for the Transvaal Clothing Industry. Joseph engaged in her first organized political activity during this period, participating in a Torch Commando march and working in the campaigns of a number of Garment Workers Union leaders who were running for Johannesburg City Council seats as Labour Party candidates.

When Sachs left South Africa for England, Joseph went with him but returned soon after. By this time she had come to an understanding that the crux of South Africa's problems lay in the racist nature of the society itself. She had come into contact with blacks on a basis of equality and, returning to South Africa, she sought a more directly political home in a non-racial context. Joseph found a political home that offered both political satisfaction and direct contact with blacks in the Congress of Democrats, where she served as national vice-chairman and as secretary of the Federation of South African Women (FSAW). Joseph was a highly visible member of the Congress Alliance, playing important roles in organizing the Congress of the People and the women's protest marches on Pretoria. A defendant in the Treason Trial, Joseph was one of the group of thirty defendants who were acquitted only after years at trial. By the time of the Treason Trial, Joseph was a full-fledged member of the national liberation movement; she had spent time in jail. In October 1962 she became the first South African to face house arrest and she lived that way for the next nine years. Joseph remained an active participant in the national liberation movement, and when her banning ended, gave speeches and counseled university students on the history of the Congress Alliance. In 1983 she attended the founding conference of the United Democratic Front (UDF) and was introduced as " 'the mother of the struggle.' "[82] The South African anti-apartheid newspaper *Weekly Mail* ran an article memorializing Joseph shortly after her death, and the tone of the article is interesting. Joseph is lauded for all those characteristics one would have expected: her spirit of defiance; her willingness to consort openly with the ANC; her stature as "one of the most visible symbols of non-racialism;" her courage through bannings and house arrests. Yet there is also a certain recognition of a duality of character mingled with the positives:

> Her refusal to use her British birthright to return to what was once home, certainly served as an inspiration that helped prompt her social concern; the stature and sense of her own position that helped her stand up to the state (but also made it difficult for her comrades to oppose her); and the domineering spirit that ensured that things were done efficiently and correctly (but also did little to 'empower' those she worked with).[83]

Perhaps these mingled criticisms were directed specifically at Helen Joseph the individual. It is also possible to read them as speaking more generally to the nature of white activists of that generation, who enjoyed a greater degree of self-possession by virtue of their whiteness and so were emboldened to openly defy the state yet also may have tended to dominate discussions and push their points of view. Or, as individuals steeped in the dominant white bloc's discipline and sense of order and organization, white radicals' ability to get things done may have been coupled with a tendency always to place ends before means, insufficiently sharing the lessons of the process with those who aspired to leadership themselves. If these criticisms are valid, then Helen Joseph, one of the most selfless and influential whites active in the movement in the twentieth century, embodied the contradictions in microcosm that would make the creation of a non-racial liberation movement, and a non-racial society, such excruciatingly difficult goals to attain.

Trevor Huddleston

If one were to choose an individual who stood out, in the minds of the African people, as the living symbol of a real, lived non-racialism, who could set the example for bridging the barriers separating South Africans, that that individual could be Trevor Huddleston. Ironically, Huddleston was neither a

South African by birth nor a "people's hero" in the popular political sense of the term, as one might have described a Bram Fischer or a Ruth First. Yet Huddleston became South African, identifying intimately with both the day-to-day struggles and the greater aspirations of the African people, and in his capacity as a religious figure, a community organizer, a social and political advocate, a mentor. As a human being, he managed to cross the frontier and become one with the people of Sophiatown, the famous and infamous township outside Johannesburg and a political and cultural center in the 1940s and 1950s. In this way, he was able to incarnate, beyond simple categorization, the almost never-glimpsed possibilities of non-racialism, of living with and creating full and rich relationships among people that transcended the consciousness of race and position. His was, perhaps, a unique experience.

Trevor Huddleston was born in 1913 in Bedford, England, and graduated from Oxford University and Wells Theological College. In 1937 he was ordained and became a priest in the Community of the Resurrection (CR), Church of England. Huddleston left England for South Africa in 1943, where he took up the position of priest-in-charge at the Community's mission in Sophiatown. Huddleston, writing of his time in South Africa, would apologize for his initial apathy, which kept him from immediately involving himself in the pro-African political work of Michael Scott. Scott was a British clergyman who had also arrived in South Africa in 1943 and who took up residence in the Tobruk shantytown outside Johannesburg. However, as Charles Hooper notes, Huddleston was not apathetic for long, given Sophiatown's effect upon him:

> He was 30 when he went to South Africa, a young priest, disciplined and tempered by his CR training, in a sense still 'unused.' Fr. Raynes sent him not only to South Africa, but to Sophiatown. What he walked into was the new emerging urban Africa -- vigorous, chaotic, uninhibited, resilient, vulnerable, communal, vivid and alive; and expressing above all that untranslatable quality, *ubuntu* -- being authentic *people* is what life is all about. With his great capacity for affection, Huddleston reveled in it and responded with passion to its poverty and its riches. Huddleston found Huddleston waiting for him: he learned new dimensions of human love, out of which grew increasing commitment and an expanding identity. 'Made in Sophiatown' is still branded on his personality today.[84]

Trevor Huddleston was in a unique position among whites in South Africa who chose to identify with black aspirations, for he lived among the members of South Africa's most vibrant African community, day in and day out, for over twelve years. Over time, he ceased to be a foreigner and became a neighbor or a father or a brother. He became an authentic, full-bodied human being to the people he served, and he, in turn, came to know the people not as an objectified mass, but as individuals and equals. In his writings at the end of his tenure there, what arises time after time in Huddleston's memory is the sense of belonging that he felt, one of the "intangible things that are known only to us who have lived in places like Sophiatown."[85] What Huddleston received in Sophiatown was, perhaps, intangible, but it was clearly also powerful, and he met the embrace he received in Sophiatown with his own, measure for measure:

> There was a moment actually when I did make a conscious decision of identification with the ANC. That was after about three years, I can't remember the exact date, but I can remember the actual moment, I went down to a big meeting in Johannesburg and it was that meeting of the ANC at which I declared myself as it were, I identified with them.[86]

Huddleston not only identified with the ANC and spoke at Congress events, but he was awarded the Alliance's highest honor, that of *Isitwalandwe*, at the Congress of the People in 1955. Why did Huddleston, not a formal member of any of the Alliance's component organizations, receive this honor? Huddleston was a forthright speaker, not only against white supremacy but as a thorn in the side of both the South African and international religious community. Huddleston felt the urgency of the situation, and he spoke to that urgency. He did not shy from politics and expected the church to be no less outspoken: "If the church refuses to accept responsibility in the political sphere as well as in the strictly theological sphere, then she is guilty of betraying the very foundation of her faith: the Incarnation."[87]

Perhaps it was Huddleston's consistency that made him such an irresistible force. He could move within and between Johannesburg and Sophiatown, nationalist politics and the women's auxiliary, without alienating either party. His willingness to speak out and his ability to connect with diverse groups made him a natural spokesperson and fund-raiser. He described publicity as his "weapon" and used his press connections to focus the attention of the white public on the injustices and deprivations that were manifest in the ghettos known as "locations."[88] His ability to cajole and stump for the people of Sophiatown led to a number of successes in the material sphere: the acquisition of an olympic-size public swimming pool for Sophiatown; the initiation of the African Children's Feeding Scheme; the acquisition of instruments for Sophiatown youngsters who ultimately comprised the Huddleston Jazz Band. Less tangible, though no less appreciated, was his work in conjunction with the Congress Alliance. A key government initiative in the first half of the 1950s was the creation of a curriculum for African schoolchildren that would consciously educate them for subservient roles in an apartheid society. At the time of the introduction of Bantu Education, and the boycott of same led by the African National Congress, Huddleston was instrumental in the attempt to set up alternative schools, known as "culture clubs." If he was not a member of the African National Congress or the Congress of Democrats, he lived at the heart of the battles of the period nonetheless, interacting with communists and

African nationalists and government officials, always maintaining his clarity of goals and personal integrity.

Huddleston's departure from South Africa in 1956 was at the order of his Community, and admittedly tested his obedience to its fullest extent. He returned to England and there helped found the British Anti-Apartheid Movement. He returned to Africa as Bishop of Masasi in Tanzania and later held the position of Bishop of the Indian Ocean in Mauritius before returning to England. Still, having taken out South African citizenship in the 1950s so as to be able to speak with greater authority, Huddleston never stopped being a South African, and in 1990 returned to South Africa as a full delegate to the African National Congress's first post-exile national conference.

Helen Joseph and Trevor Huddleston suggest the eclectic nature of the white radical population: It is doubtful that they considered themselves "radical" in any programmatic sense, though they recognized that their beliefs and the stands they took placed them at the margins of white society. If there is a commonality that binds them, it is the process by which each was revealed to him- or herself in the process of serving others. For Helen Joseph it was work with the Cape Flats community which led to tremendous feelings of acceptance and revelations of common humanity. Much the same is true for Trevor Huddleston, who found a life's mission in Sophiatown, among people who made him alive to the possibilities of love and respect transcending barriers. The great majority of those discussed here sought to play some part in the political whirlwinds whipping across South Africa in the 1950s, and most did so as members or allies of the South African Congress of Democrats.

Notes

1. "Where Do We Go From Here?" (speech), Piet Beyleveld, president, Congress of Democrats, at 1958 National Conference, COD Microfilm, CAMP Microfilm 1671, Section 13, 1958 National Conference.
2. Interview, Denis Goldberg, London, 3 February 1987.
3. Gustav Saron and Louis Hotz, *The Jews in South Africa: A History* (Cape Town: Oxford University Press, 1955), 2-3.
4. Ibid, 14.
5. Marcus Arkin, ed., *South African Jewry: A Contemporary Survey* (Cape Town: Oxford University Press, 1984), 3.
6. Ibid.
7. Saron and Hotz, *The Jews in South Africa*, 60.
8. Gideon Shimoni, *Jews and Zionism: The South African Experience (1910-1967)* (Cape Town: Oxford University Press, 1980), 7.
9. Interview, Michael Picardie, Cardiff, 24 January 1987.
10. Ibid.
11. Shimoni, *Jews and Zionism*, 10.

12. Interview, Wolf Kodesh, London, 12 February 1987.

13. Shimoni, *Jews and Zionism*, 7.

14. Arkin, *South African Jewry*, 5-6.

15. T. Adler, "History of Jewish Workers Clubs" in *Papers Presented at the African Studies Seminar, University of the Witwatersrand* (Johannesburg: African Studies Institute, 1977), 5.

16. Ibid., 7.

17. Ibid., 8-9.

18. Ibid., 12-13.

19. Ibid., 19.

20. Ibid., 23.

21. Ibid., 25.

22. Ibid., 31.

23. *Forward*, 12 May 1933.

24. Gustav Saron, "The Making of South African Jewry" in Leon Feldberg, ed., *South African Jewry (* Johannesburg: Fieldhill Publishing Co. Ltd., 1965).

25. Adler, "History of Jewish Workers Clubs," 44.

26. Interview, Ben Turok, London, 27 January 1987.

27. Ibid.

28. Ibid.

29. Ibid.

30. Ibid.

31. Interview, Denis Goldberg, London, 3 February 1987.

32. Ibid.

33. Interview, Amy Thornton, conducted by Julie Frederikse, held in the South African History Archive, Johannesburg.

34. Interview, Alan Lipman and Beata Lipman, Cardiff, 25 January 1987.

35. Eddie Roux, *Time Longer than Rope: The Black Man's Struggle for Freedom in South Africa* (Madison: University of Wisconsin Press, 1948), 148.

36. The South African Jewish Board of Deputies, Resolution 24, Public relations session, Twentieth Congress, Johannesburg, September 2-5, 1955, quoted in Arkin, *South African Jewry*, 13.

37. Shimoni, *Jews and Zionism*, 279.

38. Ibid., 280.

39. Interview, Hymie Barsel, conducted by Julie Frederikse, held in the South African History Archive, Johannesburg. Joffe is also mentioned in a reminiscence in T. Adler, "History of the Jewish Workers Clubs" in *Papers Presented at the African Studies Seminar, University of the Witwatersrand*, 1977.

40. Interview, Hymie Barsel, conducted by Julie Frederikse, held in the South African History Archive, Johannesburg.

41. Simons, 537. The author remembers attempting to sell the paper at a Day of the Covenant celebration at a kopje near Melville. He was able to sell nearly 100 copies of the paper before a group of women dressed in traditional Afrikaner costume picked him up and hurled him down the kopje, saving him from a beating at the hands of the large crowd of Afrikaner men.

42. Jessie McPherson, Mayor of Johannesburg and a member of the Labour Party.

43. Interview, Hymie Barsel, conducted by Julie Frederikse, held in the South African

History Archive, Johannesburg.

44. *What is Zionist Socialism?*, South African Zionist-Socialist Party (1944), quoted in Shimoni, *Jews and Zionism*, 189.

45. Lecture, Dr. Ellen Hellman, 1944, as quoted in Shimoni, *Jews and Zionism*, 188.

46. Ibid.

47. Interview, Michael Picardie, Cardiff, 24 January 1987.

48. Ibid.

49. Interview, Denis Goldberg, London, 3 February 1987.

50. Interview, Norman Levy, London, 10 January 1987.

51. Among those she remembers are Albie Sachs, Ben and Mary Turok, Denis and Esme Goldberg, and the Festensteins.

52. Letter, *Transvaler*, 11 September 1956, in Shimoni, *Jews and Zionism*, 228.

53. Interviews, Denis Goldberg, London, 3 February 1987 and Michael Picardie, Cardiff, 24 January 1987.

54. Naomi Mitchison, *A Life for Africa: The Story of Bram Fischer* (London: Merlin Press, 1973), 26-30.

55. Mary Benson, "A True Afrikaner," *Granta*, 19 (Summer 1986): 198-223.

56. Ibid, 208.

57. Benson, "A True Afrikaner," 207.

58. Ibid., 32-33.

59. Thomas Karis and Gail M. Gerhart, *From Protest to Challenge: A Documentary History of African Politics in South Africa, 1882-1964*, Vol. 3, *Challenge and Violence, 1953-1964* (Stanford: Hoover Institute Press, 1977), 29.

60. Ibid., 201.

61. Ibid., 208.

62. Mitchison, *A Life for Africa*, 155.

63. Ibid., 168.

64. Joseph, 160.

65. A vivid portrayal of this aspect of the Fischer existence in this guise is presented through the portrayal of Lionel Burger in Nadine Gordimer's *Burger's Daughter* (Middlesex: Penguin, 1979).

66. *Contact*, 8 March 1958.

67. Karis, Carter and Gerhart, Vol. 3, *Challenge and Violence, 1953-1964*, 7.

68. Benson, "A True Afrikaner," 200.

69. Helen Joseph, *Side by Side* (London: Zed Press, 1986), 158-9.

70. For a discussion of Sachs's life and autobiographical sketches of some of the leading organizers who worked with him, see E.S. Sachs, *Rebel's Daughters* (Alva: Robert Cunningham & Sons, Ltd., 1957).

71. Simons and Simons, *Class and Colour*, 324-325. For a discussion of Roux's early life and relationship with radical politics and the Communist Party see Eddie and Win Roux, *Rebel Pity* (London: Rex Collings, 1970).

72. Simons and Simons, *Class and Colour*, 538.

73. Ibid., 470.

74. See Sachs, *Rebels Daughters*, 40-44.

75. Delegations with Afrikaner components visited the Soviet Union annually between 1933 and 1938, and spoke of the excellent conditions for workers there, and against the fascist danger. See Simons and Simons, *Class and Colour*, 470-471.

76. Interview, Bettie du Toit, London, 11 June 1988.
77. Ibid.
78. Joseph, *Side by Side*, 88.
79. Ibid., 24-25.
80. Ibid., 27.
81. Ibid.
82. Ben Turok, "White Revolutionaries in South Africa," *Southern Africa Review of Books*, Vol. 1, No. 1, July 1987, 17.
83. *Weekly Mail*, Vol. 9, No. 1, January 8-14, 1993, 8.
84. Charles Hooper, "Isitwalandwe," *Southern Africa Review of Books*, Vol. 2, No. 1, Oct.-Nov. 1988, 8.
85. Trevor Huddleston, *Naught for Your Comfort* (London: Collins, 1956), 110.
86. Interview, Trevor Huddleston, London, 27 February 1987.
87. Huddleston, *Naught*, 232.
88. Ibid., 32-33.

6

Organization and Intellectual Life in the Congress of Democrats

The Congress of Democrats (COD) can only be understood in the context of the Congress Alliance as a whole. COD existed because the African National Congress (ANC) and the South African Indian Congress (SAIC) willed that it be so. Yet those leading black organizations could exercise little control in shaping COD's membership and, therefore, its ideological bent or particular activist tendencies.

The initiative taken by the ANC in inviting whites to form an organization implied certain fundamental parameters in COD's relationship with the ANC, and with the Congress Alliance as a whole. COD was to subordinate itself to, and take its lead from, the African National Congress. The ANC, as the representative of the majority of the South African population, in a predominantly African nationalist struggle, was at the helm of the Congress Alliance. The Alliance was organized along racial or ethnic lines for specific reasons. Inherent in this form of organization was the belief that each group was in the best position to organize among its natural constituency: Africans in African areas, Indians in Indian areas. COD, a white organization, would bring the Congress message to white South Africans.

COD was the unknown factor in the Congress equation. The ANC, Indian, and Coloured Congresses were immediately comprehensible as representatives of the oppressed. COD could make no such claim, and the motivations of its members were subject to inquiry and speculation. This was so not just because COD members were white and thus anomalous. It was generally recognized that when the Darragh Hall meeting was held to form a white component organization in the Congress Alliance, with liberals and radicals in attendance, the liberals opted out of any possible alliance with radicals: specifically, with members of the Communist Party.

Indeed, there were many members of the South African Communist Party in COD; they comprised much of the COD leadership.[1] The combination of "white and communist" raised the specter of a dual agenda, not just in the minds of external Congress critics, but among some Congress members as well. As members of the underground South African Communist Party (SACP) from 1953, communists could be expected to maintain allegiance to the party, its ideological precepts and long-term goals. It was plausible to ask whether white communists might not be more interested in furthering the aims of the Soviet Union than pushing forward the national liberatory agenda of Africans as subordinates of the African National Congress. In a more general vein, the "white communist" label raised a question of discipline. Could these people be expected to subordinate their interests, their opinions, their sense of self-assurance, and drive to the dictates of the ANC? Could COD, a conglomerate of individuals who enjoyed the privileges of white society and were used to getting their own way therein, discipline itself?

The answer undoubtedly varied over time and space, and was dependent in part upon the size of a given COD branch and the will of its membership, the strength and degree of organization among the other Congress affiliates in a given neighborhood or city, and COD members' understanding of their role in the Alliance and in the national liberation movement generally. The tensions inherent in the very presence of whites, organized in an apparently powerful sister organization, could easily be fanned if COD members were seen to be overstepping the bounds of their mandate.

The fear that white communists would attempt to hijack the Alliance by injecting Alliance deliberations and pronouncements with Marxist rhetoric and ideas was a live one, particularly among some members of the African National Congress. Quite a few Congress of Democrats members participated in the active intellectual life of the discussion clubs in the 1950s, and in some loose sense constituted themselves as a kind of unofficial intelligentsia of the Congress Alliance.

Discussion club members saw themselves as participants in something akin to laboratories of struggle: A great deal of emphasis was placed on theoretical work. Though the discussion clubs were absolutely separate from the Congress Alliance affiliates, members of the Congress of Democrats did at times attempt to bridge the worlds of the discussion clubs and the Alliance partners, and this raised serious questions of trust, loyalty, and spheres of influence within the Alliance. These concerns would dog the Congress of Democrats to varying degrees throughout the life of the organization. Such concerns served to highlight the tensions that were a concomitant of white participation in a movement that had not clearly defined its understanding or vision of the nature of the battle before it.

Taking the Measure of COD

The Congress of Democrats was an urban phenomenon. The majority of COD members were in Johannesburg and Cape Town. Durban also had some organized COD activity. COD's building blocks were the local branches, constituted when five or more members lived within a convenient distance of one another.[2] Regions comprised three or more branches in proximity, which would elect regional committees. Johannesburg and Cape Town each had a regional committee during some portion of COD's existence. The Cape Town regional committee ceased to exist during the first half of 1955.[3] The national council, consisting of the national executive committee -- a body of twelve members, the president, vice-president and secretary -- and twenty further members elected by the regions oversaw the organization as a whole. Decisions pertaining to "the policy, programme, and functions of the Association" were to be made during the annual national conference, and office bearers were to be elected there. COD's founders had produced a Constitution for an organization seeking to evolve in an orderly manner with a clear hierarchy and channels of authority. However, the conditions COD membership and leadership faced would militate against such an orderly process.

Recruitment of Members

The question of how COD might gain and maintain adherents exercised the COD leadership throughout the organization's life. COD's efforts were bounded by a contradiction that could not be resolved: COD was an organization seeking to increase its membership base through work in a community that was overwhelmingly hostile to its philosophy and aims. The notion of recruitment was therefore something of a misnomer. COD carried on with its proselytizing, including door-to-door canvassing in white neighborhoods, and produced its leaflets and pamphlets aimed at whites. While these activities fulfilled COD's obligations in its chartered role as a Congress Alliance affiliate, they generally had little effect on South Africa's urban white population.

The "white problem" was conceived of in two parts. COD leadership believed that just beyond the pale of the organization lay a pool of unorganized progressives sympathetic to COD's mission. It was hoped that these people could be brought in to bolster membership and expand COD's proselytization program. Even through the government repression of the mid-1950s, COD members continued to express optimism that disaffected United Party and Liberal Party members might find their way into COD, which offered "the only home to those who believed in rights for all sections."[4] The difficulties of COD's appointed task were most starkly evident in attempts to explain to whites what they stood to gain in the transformation to a democratic and non-racial state:

> The obstacle to [whites] coming closer to our viewpoint is often fear of Black domination. Our aim should therefore be to show these people that their fears could be groundless, and that a multi-racial society is possible....We should point out that we see as little need for the Afrikaans people to fear for the future of their language and customs and identity as there is for the Xosa [sic] people. In general, there is a great need for us to emphasize that it is not the desire of Africans to revenge themselves and do unto whites what has been done unto them.[5]

Enlisting the aid of the ANC in campaigning among whites was one suggestion raised in the draft proposal. Those who made the suggestion said that the reasonableness and inclusiveness of the ANC's nationalism would alleviate white fears of being "swamped." Yet one would be hard pressed to believe that COD members envisioned Afrikaners voluntarily listening to black men propound their vision of a common and equal future. In fact, members of COD, speaking publicly, would acknowledge that the short-term interests of whites and blacks were in obvious opposition. COD members tried to justify their belief that the goals of blacks and whites merged in the longer term. But there was little that COD could bring to bear in backing this point beyond a belief in the inevitability of their cause. COD members were not above utilizing stock liberal arguments in the service of recruitment, such as the point that apartheid inhibited general economic development in the country and white pocketbooks would stand to benefit from an open society.[6] This argument, positing a capitalist outcome for the national liberation movement, did not come naturally to white radicals, and there was a certain distaste in appealing to this vision.

Regional Strength

Records of COD's work in Johannesburg suggest that what successful recruitment did take place was often a matter of word-of-mouth, or arose from a context in which COD members worked with other whites who had already evinced interest in anti-apartheid work or the organization itself. An annual conference report from the Johannesburg region in 1955 mentioned that a number of fruitful contacts had been made through the campaign against the forced removals of residents from the townships in the Western Areas, including Sophiatown.[7] COD also maintained a fairly vibrant youth branch at the University of the Witwatersrand.[8] The same report stated that the Bellevue branch was the strongest, best functioning branch in Johannesburg, and that the majority of Bellevue's membership had been recruited through discussion clubs. This is a sensible correlation, as a branch almost wholly composed of members who sought debate on issues of national liberation might have been expected to act as an energetic and committed group. The fact that Johannesburg was at the center of Alliance politics allowed COD branches to draw on this already-present level of activity, which provided the basis for much of COD's educational, fundraising, and petition-related activities.

The Western Cape was a different matter entirely. Defined by the government as an area of "Coloured labour preference," Africans' ability to find work and survive in the Cape Town environs was severely restricted.[9] With no Indian population to speak of, there was no Indian Congress affiliate in Cape Town. Even Coloured political activity was highly fragmented, with the Unity Movement, other socialist splinter groups, and the South African Coloured People's Organization (SACPO) vying for adherents. The somewhat ambiguous status of the Congress Alliance in Cape Town, with a weak ANC and SACPO, created conditions in which frustrations and misunderstandings could arise among the Congresses. COD sometimes took upon itself what appeared to be a directing role, which was not necessarily in the long-term interests of the Congress Alliance. Joe Matthews, a leading ANC member at the time, alluded to this after the fact:

> Cape Town is a peculiarity. The ANC was never able to create in Cape Town an outstanding leadership. The tendency of the Congresses was to allow leadership to pass into the hands of the Buntings, and so forth....Therefore, PAC propaganda could go down in Cape Town (that is, that the ANC was in the hands of the whites)....People knew that persons like Simons, and others, were the directing force. Non-Africans made mistakes and made their African colleagues make mistakes....Sam Kahn, Bunting and others were recognized as being important.[10]

COD members chose to explain the relative weakness of their presence in Cape Town in two ways. The first mode of explanation built upon the "weak Congress" theme. Correspondence between the COD leadership in Johannesburg and Cape Town suggests that the Cape COD leadership felt frustrated because the other Congress affiliates were not "working correctly."[11] To this complaint the Johannesburg COD respondent opined that "each congress must solve its own organizational weakness."[12] Yet, there was also self-criticism pertaining to keeping COD's own house in order in Cape Town. An internal Cape Town report described COD members as aloof, both to non-Congress organizations and toward some of their own members. Meetings were criticized as too often "cold," making little provision for newcomers and ignoring their apprehensions about joining "an 'extremist' political group."[13] One could take this self-criticism as supporting evidence for Joe Matthews' statement that, by appearance, a small clique of activists were not only dominant in COD but were the major Congress Alliance presence in the Western Cape.

COD did have a presence in Durban, but it was a small one. Durban was able to support only one branch, with a handful of members, for the majority of the Congress Alliance period. A branch report in 1954 noted four members present at a meeting; a similar report in 1955 noted six.[14] Yet, the sense in reading these early reports is that of an active branch, engaged in dialogue with other progressive organizations both within and outside the Congress Alliance. Meetings were fortnightly, and the group planned and held fund-raisers such as garage sales, film shows, and musical evenings, fairly standard fare for COD branches throughout the country. The branch also attempted to popularize some of the alternative publications produced by the leftist press, including *New Age*, *Liberation*, and *Fighting Talk*.

COD was set a difficult task by the ANC. The Alliance's message was revolutionary, and few whites proved a ready audience for it. The difficulty of stumping for the Congress Alliance among whites served only to heighten the desire among many COD members to increase the amount of their time spent working with and among blacks. While work in the white community guaranteed rebuffs, harsh words, possibly even physical violence, the COD member in a black community might expect to be received as a partner who was willing to make sacrifices in the name of justice: an extraordinary individual. Given the greater rewards accruing from work with and among blacks, it is not surprising that it was difficult at times for COD members to balance their responsibilities as Alliance ambassadors to the white population against the desire to work and find acceptance among blacks, particularly Africans. While work among whites often seemed to verge on the hopeless, work among Africans seemed vital and necessary. If liberation was to be won, it would occur through African agency, and how could time be better spent than in helping to prepare the African people for leadership in the liberation movement? The COD membership felt it had something of particular value to offer in the townships -- education. Many COD members, as members of the Communist Party, and/or with substantial formal education, were people of the book. Their world view had sprung from the printed page and they were acutely aware of its power.

COD's Educational Function and *The World We Live In*

Education was considered a matter of necessity, both within the organization and in its dealings with others. An internal COD document distributed for discussion during the run-up to the Congress of the People stated that COD needed to "sharpen the keenness" of its own members by "giving them a greater understanding of the truth of our outlook." The document suggested that political education be made a permanent feature of branch life.[15] COD never would have described itself as a vanguard organization, yet its small membership allowed it to aspire to a level of internal politicization that was out of reach of the much larger Indian or African Congresses.[16] Given the relative paucity of COD members, political education appeared to the COD leadership to be an area in which COD could make an impact upon the Alliance as a whole.

In fulfilling this self-defined role, the Congress of Democrats produced a number of lectures and notes for discussion under the titles, *The World We Live In* and *What Every Congress Member Should Know*. The former provided no formal attribution pinpointing the individual, branch, region, or executive body of COD responsible for its production. The latter was a contribution of COD's Transvaal Provincial Executive. The lectures and notes were an attempt to present a specific world view to an active or potentially mobilizable audience. *The World We Live In* attempted to explain basic Marxist and derivative economic theories and concepts, such as modes of production, surplus value, imperialism, and socialism. These concepts were then placed in the South African context and discussed with the immediate needs of the national liberation movement in mind. This context was set from the very first paragraph when the author stated that "All history is the story of the struggle of people to be free."[17]

The series began by tracing the evolution of societies and inequalities within them from the first "isolated groups," equal and without masters in their "common poverty" to societies based upon the division of labor. This leap was addressed in one paragraph, charting the progression from hunter-gatherers to settled agriculturalists and domesticators of animals, from want to surplus and ownership of property: "The old equality began to die out, the wealthier began to live not by their own work, but by the labour of the poor, who tended their animals and plowed their lands."[18] There were paragraphs on the slave system and the feudal system, and the serf was compared to a South African farm labour tenant. Feudalism gave way to capitalism, and workers were no longer tied to the land. Now they were to be tenants, "selling their hands," free yet no less impoverished than before:

> The world we live in is, then, a world divided into classes -- into masters and men. It is a world in which one small class of men, the masters, those who own the tools, the machines, the factories, the mines, the forests, the farms, live from the work of the many, the working people, who own nothing but their ability to work. This system of some living and growing rich through the work of others we [call] *exploitation*.

This first lecture was an introduction, an attempt to set up an analytical framework, in as simple and straightforward a manner as possible, before discussing the applications of these ideas to South Africa. Later lectures focused particularly on South Africa, and some included questions for group discussion. The third lecture, entitled "Change Is Needed," was a discussion of Parliament as a key instrument of the ruling class and of the institutions backing up the rule of law. To the initiate, a number of the messages in the text may have appeared somewhat ambiguous. The Parliamentary parties were presented as bankrupt institutions working to maintain the structures of apartheid. Yet it was suggested that alliances with such parties ("allies who we know will not always be with us") were not to be eschewed, and changes that enemies of the people might also want for their own reasons were to be welcomed if they made the lives of the people easier.[19] Yet having said this, the author described the structure, composed of the army, police, courts and prisons, educational institutions and media, used by the ruling class to maintain its interests, noting that:

> It is clear that such rule as this cannot be set aside by minor concessions and reforms. In the end, such a state apparatus, built upon a foundation of oppression and exploitation, can never serve the ends of the people and of the Congress movement. The Congress movement must build for itself a new kind of rule, and a new kind of state -- a state of people's equality and liberty. That kind of state we call "a People's Democracy."

The questions for discussion were not very subtly primed to generate specific ideological responses: Must the "People's Democracy" wait until the illiterate and uneducated become formally educated? Which classes would one expect to benefit from a "People's Democracy"? Who should expect to support it? Would this Democracy provide equal rights to all without exception, or "for the great majority only"? Written prior to the Congress of the People, the outlined attributes of the system resonated with the language and intent of the Freedom Charter: equal rights to vote and to be elected; nationalization of monopoly industries and mines; equitable distribution of the land among those who worked it; criminalization of all discrimination based upon race or gender; the right to form trade unions; freedom of speech, movement, and assembly; redistribution of housing to provide for those with none. The achievement of this program would depend upon taking the control of the state from the "hands of the old ruling class of exploiters" and placing it in the hands of workers and peasants, "allied with all others who see that South Africa's future happiness cannot be won while the state is the property of the exploiters and the oppressors."

The Congress of Democrats had taken upon itself the task of broadening and deepening the political educational base of the Congress Alliance's rank and file. A specific policy-oriented and ideological agenda was woven throughout these lectures; based on an orthodox Marxist approach to understanding world history generally and the history of South Africa and the national liberation movement specifically. The lectures also set out an agenda for a society in transformation, much in line with the Freedom Charter's vision but ahead of it. Necessary components of a People's Democracy were detailed and could be understood as representing a transitional form of state between the apartheid present and the Marxist-Leninist future.

Beyond the content of these pieces, the task of understanding how such lectures were used, by whom, and to what end, presents many difficulties. The lectures did circulate to the Congress affiliates prior to the Congress of the People. Mary Benson has noted of COD in the 1954-1955 period that they "circulated a series of three roneoed Marxist lectures that were sent to some ANC branches."[20] An unattributed report on the use of *The World We Live In* noted that regular study classes were being held in approximately seven areas, "mainly African areas," with an average attendance of around ten:

> The lecturers are Europeans up to now as our African comrades are somewhat reluctant to do this work. It [is] our intention however to see to it that more and more of these classes are brought into being and that Africans take a bigger share in this work.
>
> My personal experiences with the notes are that they are only a very rough guide for a class among people who are illiterate and who have no political training. The concepts of a historical evolution of various social patterns throughout the world are difficult to put across. It is difficult to be abstract or general about a society with these people. One must always refer directly to South Africa and its social set-up. For these reasons, it seems that a simpler course on the same lines is called for.[21]

The tone of the first paragraph above suggests that COD did not simply disseminate these lectures; they were clearly COD's "project," and COD was taking responsibility for the organization of the lectures in African areas and planned to intervene if possible to enlist African participation in disseminating them. It is impossible to know how many COD members, and who specifically, were involved in this project. It does appear that those involved had a specific goal in mind: to provide relatively small groupings of urban Africans with a specific Marxist historical and ideological framework from which to view the events then taking place, and to facilitate their taking leadership positions in their communities.

What did the members of COD involved in this effort know about the audience they were, in fact, reaching? What would have led these COD members to believe that this complex, class-based analysis, presented by whites from without, would find a response in African communities? The general concept of the lectures and their specific content may be most interesting for what they suggest about the consciousness of some of COD's more theoretically inclined members. There is both a definite affirmative energy and a degree of characteristic white bossiness at play in this project. The protagonists believe that the analysis they have to offer is crucial to a complete and true evaluation of the South African situation and its amelioration, and they are going to take affirmative action to share this knowledge. Yet there is also an air here of "knowing what is best for others even if those others do not see it yet," which could be the intellectual elitism of communists but which could also suggest a more general arrogance and paternalism present in white South Africans of the time.

If "African comrades" were reluctant to do this work, did their white counterparts stop and ask why that might be the case? Why did they assume that it would be possible to recruit Africans to "take a bigger share" in this work? And more particularly, why were these discussion notes geared at such a high educational level for people who "are illiterate and have no political training"?

One could suggest that this project was a measure, not only of the physical closeness in which black and white were working together and learning from one another, but of the gulf -- cultural, political, and educational -- that remained between even the most committed whites and the majority urban African population with whom they were attempting to engage. The world of theory was, by the mid-1950s, still a predominantly white world, at least in the context of the Congress movement, and the discussion clubs of urban South Africa provided a particularly rarefied atmosphere in which the revolutionary intelligentsia could meet and speak a common language.

The "Revolutionary Intelligentsia" in the Early Alliance Period

"Discussion clubs" of various types were a staple of the South African political landscape throughout much of the first half of the twentieth century. Some of the earliest clubs, like the Jewish Workers Club (JWC), served not just as venues for political discussion but as Jewish cultural centers, providing the Jewish workers of South Africa's major urban communities with social and cultural gathering places for purposes of entertainment, education, and commiseration.

Other clubs, including the Left Club, were focused to provide an arena for intellectual discourse on a broad range of topics within the ambit of the struggle for a changed society. When the Suppression of Communism Act (1950) became law, discussion of radical political and theoretical topics had, in some instances, to be carried out in non-party organizations. In the late 1940s and early 1950s there were a number of these clubs, including the Modern Youth and Modern World societies, the South Africa Club, the Forum Club, and the Johannesburg Discussion Club which provided venues for discussion of such topics.

The Johannesburg Discussion Club

The Johannesburg Discussion Club was formed in 1952[22] to provide a forum for the ongoing discussion of topics related to the struggle for the liberation of South Africa:

> The originators of the Club confidently believe that the free and frank discussion of views, based on theory and participation in the daily struggle, can only prove of inestimable value to the Liberatory struggle, and that it will directly assist in promoting its ends.[23]

The range of topics discussed is suggested by the list of papers in the Club's gazette, *Viewpoints and Perspectives.*[24] Among the topics touched upon were forms of resistance, including a critique of the Defiance Campaign; the origin and status of black trade unions; and the role of rural populations in the national liberation movement.[25] The national question continued to dominate the theoretical agenda. Among the most important sub-themes was the class nature of the constituents in the national liberation movement and the outlook specific to each. Participants constantly returned to the question of the primacy of the political or economic form of struggle at that time, i.e., whether the struggle was to assume a national-democratic or proletarian form:

> The fact that the ruling class is composed almost exclusively of Europeans, and that the Africans compose the bulk of the working class....does not effect [sic] the conclusion that this is a class society and that it is a class struggle that is being waged. What is affected by the factor of racial oppression, however, is the form that this struggle takes and the particular objectives that are set. This factor of racial oppression determines that the political struggle of the Non-Europeans shall assume a liberatory form, i.e., a struggle for full democratic rights. The economic background in its turn demands that sooner or later the movement must be grounded on working-class forms of organisation and working-class forms of struggle.[26]

This formulation was generally accepted by the Discussion Club participants, to the extent that it recognized South Africa as a society based on capitalist exploitation but a society in which the great bulk of the proletariat was African and thus doubly exploited.[27] The fact that none of the participants stated that the working class was leading the struggle or that it would take such control in the immediate future suggests that there was a general recognition of the relative weakness of organization among African workers at this time. However, discussion participants drew varied conclusions. As David Everatt has pointed out, a primary focus of Johannesburg Discussion Club dialogue was the nature of the nationalist movement and the question of whether workers' organizations should be organized as bodies distinct from national organizations. Emerging from these debates were two relatively distinct positions, one grounded in the primacy of the working class, defining South Africa as a capitalist society, the national struggle as obfuscatory, and the ANC as an organization characterized by a lack of class content or understanding. The other position, centered around the Johannesburg branch of the SACP and its theorists, revolved around those theorists' conceptualization of "colonialism of a special type" as uniquely present in the South African situation. This outlook rejected the notion that an African bourgeoisie proper existed in South Africa, took to task those who refused to engage with the active forces of liberation in the Congress Alliance, and believed that ultimately, with the integration of working-class and nationalist forces within the Alliance, the working class would come to the fore and move the struggle beyond immediate national-democratic goals.[28]

Class and the Congress Alliance

One of the most crucial points among Discussion Club members was the class composition of the Congress Alliance and the impact of that composition on the form of struggle. The analysis of a number of the membership, former or underground members of the Communist Party, was undoubtedly influenced by Soviet theory and policy pertaining to colonial or dependent nation circumstances. The Soviet period of Popular Front multi-class alliances had ended in 1947 when Zhdanov introduced his "two camps" theory. Zhdanov's theory dichotomized the world into democratic anti-imperialist and imperialist anti-democratic camps. Alliances with national bourgeoisies were to be ended, and new nations brought to independence by indigenous bourgeoisies were to be considered imperialist. Policies of non-alignment and bourgeois-led social reform were not considered valid.[29]

The policies stemming from the Zhdanov thesis proved unrewarding and insensitive to the realities of change and government among newly independent post-colonial nations. Thus there followed a reorientation of Soviet policy from the early 1950s that acknowledged decolonization as a legitimate step toward full independence. This in turn revived the virtue of pan-class alliances: "Now it was argued that the interests of the national bourgeoisie in the underdeveloped world would lay in the formation of an alliance with the working class, the petty-bourgeoisie and the peasantry, in order to struggle with them against imperialism."[30]

By the mid-1960s this line of thought had developed among Soviet theoreticians into a mutation of the classic Marxist-Leninist theory of transformation from capitalism to communism. It was now posited that in the decolonized world the process of socialist transformation was more complicated than had formerly been understood. There was a process taking place between the demise of capitalism and the achievement of socialism that was described as "national-democratic." National democracy represented a break with capitalism and was characterized by nationalization and the building up of the state's productive structure.[31]

In grappling with a theoretical understanding of the problem of South Africa's liberation, members of the party had developed the theory of internal colonialism, later described as "colonialism of a special type." The Communist Party was analyzing the South African socio-economic terrain in light of the theory of internal colonialism by the early 1950s: "In a word: there are two nations in South Africa occupying the same state, side by side in the same area. White South Africa is a semi-independent imperialist state: Black South Africa is its colony."[32] Michael Harmel, a leading party theoretician and the primary shaping force behind the theory, noted at the beginning of his 1954 Discussion Club paper that, with the exception of the Eastern Bloc and China, the world was divided into two camps, the Imperial and the Colonial, a statement mirroring straight Zhdanovian rhetoric. Harmel did not have to make mention of relations with the indigenous South African bourgeoisie because there simply was no such class in significant proportion:

> The emergence of a significant capitalist section among the Africans has been deliberately frustrated. The African liberation movement is not dominated by the unstable and potentially treacherous elements which have led similar movements elsewhere. It is a movement of workers and peasants, professional people, middle and commercial classes in which the progressive, working-class tendency plays an increasingly influential part.[33]

Peter Hudson has stated that by 1956 the dominant Soviet theoretical orthodoxy acknowledged the error made in labeling the indigenous bourgeoisie as a "potentially treacherous element."[34] Harmel's analysis incorporated aspects of both Zhdanov and the type of analysis that embraced what was defined officially in 1960 as "national democracy."[35] Working from the assumption that the national liberation movement was fighting an imperialist force, albeit an internal one, Harmel noted that while the Movement had focused on political demands, "the economic content of national liberation in South Africa must inevitably centre in redivision of the land and the nationalisation of the principal means of production (for the power of imperialism in this country can only be broken by divorcing the imperialists from the means of production)."[36] Harmel echoed the prescriptions that would become integral to the definition of the national-democratic phase of struggle as defined by Soviet theoreticians in the following decades. One might also note the apparent foreshadowing of the most controversial clauses of the Freedom Charter in Harmel's statement.

Rusty Bernstein set out much the same analysis of the bourgeoisie in his Discussion Club paper. Like Harmel, Bernstein acknowledged that there was no real black "traitor class," with the exception of a small number of individuals "so isolated, so bankrupt, so exposed that a struggle against them would not be a service to the movement."[37] However it was the very fact of the multi-class nature of the struggle and its ability to mobilize broad sections of the population that fueled the national liberation movement. The majority of those working therein belonged either to the working class, the professional stratum of doctors, teachers, lawyers, and religious leaders ("less conscious but of great influence because of their educational level and their specialized skills") or that group comprising traders and craftsmen, carpenters and hawkers, the petite bourgeoisie.

Bernstein's analysis was ambiguous as it applied to the oppressed bourgeoisie. He did not deny completely that there were members of the black bourgeoisie with an interest in the struggle. The inability of this class to realize its potential under apartheid was manifest. Rather, Bernstein saw a divided bourgeoisie, with, for example, supporters in the Indian Congress and in the Indian Organization, a conservative, anti-communist rival; supporters in the ANC, and supporters in the breakaway Thema group, a "nationalist-minded" anti-communist and anti-foreign ANC splinter group. The oppressed bourgeoisie, what there was of it, was generally and roundly condemned as an exploiting class, yet it was not necessarily seen as standing completely outside the pale of the liberation alliance.

That the working class *should* lead the struggle was taken for granted. That the working class was sufficiently conscious and organized to do so in the early 1950s was not taken for granted, nor did Bernstein take for granted that the working class knew what was to be gained following a successful struggle that put a capitalist democracy in place:

> They [the working-class] have this to win: 1. The best possible conditions for political organization. 2. Experience in political leadership of all sections of the working population. 3. The abolition of colour oppression and thus the clearing away of the race versus race issue, which will leave the class issue clear and exposed for all to see.[38]

Bernstein took those to task who saw the national liberation movement as bourgeois, anti-socialist, or reactionary. To Bernstein this was a misreading of Stalin ("the slogans of nationalism arise in the market place") and proof of an inability to understand that different classes sought different gains from a successful struggle. If there were no specific reference made, by either Harmel or Bernstein, to the emergence of a bourgeois democratic state as a transitional phase on the road to socialism, that was certainly implied by both of these discussants, and it explains much about the fervent participation of members of the Communist Party along with other progressive whites in the Congress Alliance.

It is interesting to note that in the introductory essay to the third issue of *Viewpoints and Perspectives* in February 1954 the editor raised the possibility that the indigenous bourgeoisie might become more prominent and play a larger role in the national movement:

> [The] "Discussion" editorial, on the other hand, poses the question of whether the emergence of a native bourgeoisie playing a leading role in the Liberatory movement is not a likely development. Will the development of capitalism in this country not, however, create favorable conditions for the crystallization of a native bourgeoisie?
>
> In view of the backwardness of the proletarians, the fact that they are still largely unorganized and not integrated fully with the capitalist system -- would this not give the native bourgeois and petty-bourgeois elements an opportunity to play a large part in the national movement for some time to come?[39]

Harmel and Bernstein reached no consensus as to the class nature of the national liberatory movement, beyond the acknowledgment that it was a multi-class alliance in which the progressive workers' tendency appeared to play an increasingly visible part. Danie Du Plessis, a primary dissenter (though also a party member) claimed that the leadership of the national liberatory movement was bourgeois and would forsake the interests of workers in a time of crisis to protect themselves.[40] Du Plessis stood alone in calling for the amalgamation of the Congress affiliates within a single organization such as the African National Congress.[41] Du Plessis' point-of-view can be explained by his analytical emphasis, which, unlike that of either Rusy Bernstein or Michael Harmel, understood class to be the primary factor influencing the struggle of the oppressed, so militating in favor of a struggle with worker and peasant leadership. Bernstein replied that such an organization could exist but only in the context of building socialism, which could take place only after national liberation. He thus defined the primary struggle as national in character, to be led by the primary oppressed national group, the African people, in a multi-class alliance such as Congress.

Danie du Plessis, alone, described the leadership of the movement as bourgeois. No one else could find the bourgeoisie to which he was referring. Rather, it was clear to those in the SACP that the majority of dynamic actors in the struggle at the time were intellectuals and professionals. If the discussion participants could find evidence that the black working class was expanding its participation in the political struggle and was ripe for organization, they were also aware that the struggle was hardly working-class in character. Therefore, the sentiment was expressed time and again in the dialogue that the prevailing multi-class alliance had to be accepted, with a view to working-class hegemony in future.

The National Question and "Nationalisms"

In 1954, two Cape clubs, the South African Club and the Forum Club, sponsored a "Symposium on the National Question." The papers presented in the Symposium and at least a partial transcription of the discussion afterward were printed later that year.[42] In his paper, "Nationalisms in South Africa," Lionel Forman, then editor of *New Age* (a successor to the *Guardian)* and a member of the SACP set out a number of points that he expected the club members to agree on. The first of these points was simply that "basic to everything in South Africa is the capitalist system. It is sometimes said that it is the 'racialist policies' of the Government which are basic. That is incorrect."[43] Forman's second point of consensus was the "double yoke of oppression" faced by Africans, with an African proletariat much aware of its oppression as a class and as a national group, and with an African intelligentsia and petty bourgeoisie acutely aware of its oppression on a national basis. Thus it would be necessary to delineate the nature not only of the oppressed and oppressor classes, but of the oppressed nationalities. Forman went well beyond this, however, calling for the delineation of those groups that constituted "nations" or "national groups" with an eye toward the promotion of such entities' self-determination in a "people's democracy":

> It looks like apartheid. But of course the guarantee of national autonomy in a people's democracy bears not the slightest resemblance to apartheid. Though the citizens will have their own national territory where their language and culture reigns supreme, they will be free to go and to live where they please, and where they are a substantial group outside their territory, then they too, will have language and other rights.[44]

Forman put forward his ideas with such fervor, probably because he was influenced by Soviet nationalities policy in the 1950s, policies he had the opportunity to study during time spent in eastern Europe.[45] Forman's analysis took off from that of Professor I.I. Potekhin, then director of Moscow's Africa Institute. Potekhin had declared that groups such as the Zulu and Xhosa would call for self-determination and such calls should be respected.[46] Forman was not unaware of the criticisms leveled at those who used the rhetoric of ethnic division. Yet he described as "superficial" thinking that which could not distinguish between "people's nationalism," a progressive force as witnessed in the Soviet national republics, and apartheid-based notions of ethnicity and ethnic segmentation.[47] In the Discussion Club debate, Forman stood starkly alone in claiming that the time would come when the right to self-determination would "become an urgent demand."

Jack Simons responded quickly and surely to Forman's paper, criticizing the rhetoric of cultural autonomy and self-determination and warning that any theory which talked of " 'development along own lines' " should be immediately suspect. For Simons, the only valid nationalism was that found in the multi-class national liberation movements. Simons believed that with the rise of industry, the effect of the color bar, and the concomitant blunting of class aspirations, the role of workers in those movements would increase. While not denying that workers had an awareness of and interest in ethnicity or cultural life, Simons believed this tendency would diminish in the urban areas as workers mixed freely and developed "internationalist" outlooks. Rather, the petite bourgeoisie, the teachers, ministers, journalists and professionals, with a greater material interest in the delineation and maintenance of cultural practice, would have a greater interest in maintaining cultural boundaries.[48] Simons found little interest among Africans in the ANC for any project that would divide them at a time when "[t]heir whole attention is concentrated on mastering the new environment, of which they have become a part, and of adapting their traditional life pattern to these circumstances. There is much in the old tribal culture that is a handicap and a burden."[49] The demand for the equal recognition of languages in the context of a post-apartheid, non-racial, and worker-led state was not inconsistent.

K.A. Jordaan, a leading Cape Town intellectual, immediately warned against taking up "those artificial *Herrenvolk* race divisions" as points of departure. Jordaan defined four national groups, Africans, Coloureds, Indians, and Afrikaners, none of which, he posited, was interested in existence as a physically separate nation. Jordaan noted that where tribal patterns persisted, they were "assiduously preserved by the ruling classes and combined with the most modern forms of exploitation and oppression."[50]

Jordaan did not scoff at the possibility that after the national liberation movement had reached its goals a demand for cultural autonomy could arise, and he accepted such demands as legitimate. However, Jordaan did not see that questions of cultural autonomy had to be discussed before such demands had arisen. Jordaan was a proponent of the concept of permanent revolution and believed that colonialism had destroyed most "traditional" structures. It was thus reasonable to look to a revolution leading not to national liberation in the context of a newly freed capitalism but to the complete overthrow of present capitalist structures and the constitution of socialism.[51] Whether communists or socialists, few of those in the clubs were willing to face the possible complications presented by "traditional" societies. Either such societies were deemed no longer to exist, or they could be dealt with after national liberation. Ethnicity was dismissed out of hand, although the basis for this dismissal was unclear. The Soviets had acknowledged their own ethnic problem and had granted autonomous regions to recognized ethnic populations. Perhaps there was a certain smugness among Soviet supporters at this time which led them to believe that they would treat with national ethnic problems if and when necessary.

Jordaan's analysis of the class basis of the democratic forces followed closely that of Forman and Simons. He found no oppressed bourgeoisie to speak of. The working class formed the bulk of the democratic camp. Jordaan looked beyond the struggle itself to the victory of the "democratic" forces and the creation of a capitalist-based national-democratic state. Jordaan foresaw that liberating laws which freed up land ownership and abolished the color bar would not fulfill the aspirations of the working classes but would guarantee the permanence of a revolution leading to socialist reconstruction in the context of a "tottering world capitalist system." This was the thrust of the analysis the Communist Party was promoting throughout the 1950s and early 1960s. However Jordaan was willing to state in a public forum what most party members would not make explicit, that is, the transitional nature of the victory of national democracy itself.

The clubs were "think tanks" for the revolution where the views discussed here and many more were exchanged. Despite many differences of opinion, there were a number of points of general agreement. They agreed that the struggle facing the oppressed had to be understood in class terms. South Africa was a capitalist society, and the structures of race effected by the white minority with the complicity of international capital served the purposes of accumulation by that minority. If the nature of the struggle at that moment took on a political form against an apparently racial political structure, both had to be understood in light of the form taken by the class struggle at that time. The demand for political revolution and the creation of a democratic capitalist state was but an interim stage on the road to the victory of the proletariat and the building of socialism. The struggle did not pit black against white, but, as Jordaan put it, the struggle had "a place for every person who subscribes to its programme. It follows therefore that there is a place for every South African in the society of the future."

White democrats believed they had a rightful and unashamed place in the struggle. Eddie Roux spoke of the debt owed by the African nationalist movement "to left wing groups and parties initiated by Europeans in this country, particularly the Communist Party," and of the role of the *Guardian* ("in spite of many tactical vacillations") in raising "the political level of the Non-European people."[52]

Harmel, in his analysis of imperialism and South Africa, stated that if the administrators and policemen of apartheid were to reject those roles in which they had been cast "by the Chamber of Mines millionaires," one of the factors that would lead them to do so would be "the effectiveness and perseverance of that clear headed and courageous band of white democrats which has already identified itself with the aims and the struggle of the liberation movement."[53] Claims such as this could more easily be made in the rarefied atmosphere of the discussion clubs. However, the intellectual armor of the white democratic group could not insulate it from the realities of struggle in the townships or in the white suburbs and working-class communities, and that reality was generally more uneven and complex than was accounted for in theories of class formation.

Conclusion

COD was chartered by the ANC to spread the Congress Alliance message to the white population of South Africa's urban areas and to win converts to the national liberation movement, if possible. COD's structure was such that all levels of the organization would maintain contact with their partners in the other Alliance affiliates, while the building blocks of the organization, the local branches, would be grounded in white neighborhoods into which COD members would move, bringing the Congress message.

A significant number of COD members had also been members of the Communist Party (and, with the Party's reconstitution in 1953, probably were members of the underground South African Communist Party) and would continue to make manifest their belief in the primacy of class struggle. In doing so, they did not necessarily represent the COD leadership or COD's broader membership. Nor were COD members moving into white areas to continue the work the party had begun in attempting to capture the white working class. COD members working in white areas were carriers of a more specific and undoubtedly more controversial message: that apartheid had to end, and the black majority had to be given its basic human and democratic rights. There was little, rhetorically, that COD members could offer the white on the street in putting forward this message. Moral suasion was not particularly effective. Raising visions of cataclysmic violence as the end result of white greed may have moved some of those approached to fearfulness, but, again, it was a doubtful tactic. Appeals to white self-interest, while a more realistic tack, were ineffective, because while the COD argument suggested whites would prosper increasingly in a free society, whites were prospering quite nicely under apartheid.

The difficulty of working among whites was undoubtedly one force moving COD members to work with blacks in black areas. Exhortations were made from time to time in an attempt to re-energize members for work in white areas, and it was stressed that there was no alternative to "slow, slogging and tenacious political work among our mentally depressed brethren, the Europeans" if the Nationalists were to be overcome.[54] Yet COD members did find alternatives. The political work of the white suburbs was often stultifying, but work with and among blacks was just the opposite, as it affirmed the relationships on a microcosmic level that COD members sought to create generally. However, the relationships forged by COD members and branches from area to area and over time were subject to all the normal stresses of political work. And these stresses were compounded by the difficulties and potential misunderstandings that could arise from the dynamics of race and class, the effects of which even the most advanced and conscious Alliance members were subject to in the South Africa of the 1950s.

The whites who formed the Congress of Democrats were a rare group. They had placed themselves publicly in alignment with the cause of African liberation and the destruction of the apartheid system. They were an eclectic group, encompassing individuals like Helen Joseph, who was experiencing only her second organized political experience as a COD leader, and Rusty Bernstein who had been involved in organized, radical political work since he was a teenager and wrote discussion papers on questions of Marxist theory. What they shared was a common abhorrence of white supremacy and a desire to identify with, and to be seen to have identified with, the forces ranged against the government and white society. Some were moved by visions of non-racial worker mobilization and socialist revolution, while others were moved by their revulsion at the scars of oppression and the possibilities of creating a common culture in a democratic society.

Notes

1. It should be noted that not all members of the former CPSA, or of the SACP constituted in 1953, joined COD. Among those not joining were Jack Simons and Ray Alexander. One might speculate that a number of the communists who did not join COD objected to the racial form of organization within the Congress Alliance. In a conversation with Jack Simons, he alluded to COD's creation along "racial" lines. Conversation with Jack Simons, Lusaka, 24 August 1988.

2. Congress of Democrats, Constitution, 3.

3. Congress of Democrats, Draft Organizational Report, Annual General Meeting, September 1956, Treason Trial Documents, Cooperative Africana Microform Project (CAMP) Microfilm 405, Reel 14.

4. Congress of Democrats, Organizational Report, Minutes, National Conference, Johannesburg, 24 June 1955.

5. Congress of Democrats, draft statement, "What is Our Perspective?," n.d., South African Congress of Democrats microfilm, CAMP Microfilm 1671, Section One, Minutes and Correspondence. Note that the term "multi-racial" here does connote a non-segmented post-apartheid society.

6. Congress of Democrats, "Speaker's Notes," n.d., n.a., COD Microfilm, CAMP Microfilm 1671, Section Ten, Statements and Letters.

7. Congress of Democrats, Minutes, National Conference, Johannesburg, 24 June 1955, "Organisational and Financial Report," Treason Trial Documents, CAMP Microfilm 405, Reel 22.

8. Springbok Legion/Congress of Democrats, Executive Committee Report, 30 September 1953, Treason Trial Documents, CAMP Microfilm 405, Reel 11, and *CounterAttack: Bulletin of the South African Congress of Democrats*, 1956, Treason

Trial Documents, CAMP Microfilm 405, Reel 3. The September 1953 report noted a combined membership in the then joint SL/COD branches in the Johannesburg area of 123.

9. See Tom Lodge, *Black Politics in South Africa Since 1945* (London: Longman, 1983), 214, for the weakness of the ANC in the Western Cape. For a discussion of the government's labor preference policies as an aspect of apartheid restructuring see Ian Goldin, "The reconstitution of Coloured identity in the Western Cape" in Shula Marks and Stanley Trapido, *The Politics of Race, Class and Nationalism in Twentieth Century South Africa* (London: Longman, 1987), 156-181.

10. Gwendolen Carter Papers, Northwestern University Archives, Box 39, File 7. T. Karis interview with Joe Matthews, summer 1963.

11. Carter-Karis papers, Document F39:1-6, Letter J. 19.

12. Ibid.

13. Congress of Democrats, Cape Town Draft Organisational Report for Annual General Meeting, September 1956.

14. Congress of Democrats, Durban Branch, minutes of branch meetings held on 20 December 1954 and 17 August 1955, Treason Trial Documents, CAMP Microfilm 405, Reel 12. Perhaps the most well-known members of this small group were Rowley and Jacqueline Arenstein. Rowley Arenstein was a prominent Durban attorney and a member of the Communist Party of South Africa from 1938 through the later 1940s. Jacqueline was a defendant in the Treason Trial.

15. Congress of Democrats, "Notes on the Political Situation for Discussion at Conference," National Executive Committee, 1955, Treason Trial Documents, CAMP Microfilm 405, Reel 5.

16. Jack Hodgson, "Draft of the Immediate Program of Action": "c.") "Education of Members. It is proposed: I) that prepared lectures and discussion notes on appropriate subjects should be furnished by the National Council. II) that panels of speakers and lectures should be organised by the National Council, Regional Committees and that branches that are too remotely situated to avail themselves of these should supplement the prepared lectures and speakers notes by inviting appropriate people in their particular community to address them on suitable subjects."

17. *The World We Live In*, n.d., n.a., Treason Trial Documents, CAMP Microfilm 405, Reel 1.

18. Ibid.

19. *The World We Live In*, Lecture 3, "Change is Needed."

20. Mary Benson, *The African Patriots* [typescript], Ch. 20, 470, School of Oriental and African Studies [Ms. 348942]. Benson states that the lectures went "untranslated and uncirculated," ending up in the files of the Security Branch. That was not strictly the case, given the evidence that follows.

21. "Report on the Use of the Notes, 'The World We Live In,' " n.d., n.a., Treason Trial Documents, CAMP Microfilm 405, Reel 7.

22. Baruch Hirson states that the Johannesburg Discussion Club was formed by "dissident" Communist Party members. Hirson says that the Club failed because a number of organizations, invited to participate, refused, including the Progressive Forum.

23. Johannesburg Discussion Club, *Viewpoints and Perspectives*, Vol. 1, No. 1, March 1953, Treason Trial Documents, CAMP Microfilm 405, Reels 7 & 20.

24. This discussion is based upon review of those issues available to me on microfilm, a limited number of issues. I do not know when the journal ceased publication, but it is possible that the issues cited constitute a minority sampling.

25. E. Roux, "Active and Passive Resistance: A Study in Political Methods with Relation to South Africa," D. Tloome, "The Origin and Development of Non-European Trade Unions," Z. Sanders, "Aspects of the Rural Problem in South Africa," Treason Trial Documents, CAMP Microfilm 405, Reel 15.

26. *Viewpoints and Perspectives*, Vol. 1, No. 1, March 1953. The quoted comment was made by Michael Hathorn in the discussion following E. Roux's paper on the Defiance Campaign and political methods.

27. Eddie Roux appears to have been the sole noted participant in the Discussion Club who did not agree that South Africa was a capitalist society, saying only that in South Africa one found "an imperial and colonial relationship coexisting in the same country."

28. David Everatt, "Alliance Politics of a Special Type: The Roots of the ANC/SACP Alliance, 1950-1954," *Journal of Southern African Studies*, Vol. 18, No. 1, March 1991, 19-39.

29. Peter Hudson, "Images of the Future and Strategies in the Present: The Freedom Charter and the South African Left," in *South Africa Review 3*, (Johannesburg: Ravan, 1985), 262.

30. Ibid., 263.

31. Ibid.

32. M. Harmel, "Observations on Certain Aspects of Imperialism in South Africa," *Viewpoints and Perspectives*, Vol. 1, No. 3, February 1954, Treason Trial Documents, CAMP Microfilm 405, Reel 15.

33. Ibid.

34. Hudson, "Images of the Future," 263.

35. Ibid., 265.

36. Harmel, *Viewpoints and Perspectives*, Vol. 1, No. 3.

37. L. Bernstein, "The Role of the Bourgeoisie in the Liberatory Struggle," *Viewpoints and Perspectives*, Vol. 1, No. 2, January 1953, Treason Trial Documents, CAMP Microfilm 405, Reel 15.

38. Ibid.

39. Introductory essay, *Viewpoints and Perspectives*, Vol. 1, No. 3, February 1954, Treason Trial Documents, CAMP Microfilm 405, Reel 15.

40. Danie du Plessis, "The Situation in South Africa Today," Vol. 1, No. 3, *Viewpoints and Perspectives*, February 1954, Treason Trial Documents, CAMP Microfilm 405, Reel 15.

41. The call for "One Congress" was repeated throughout the 1950s and will be discussed below.

42. "Symposium of the National Question," Liaison Committee of the South Africa Club and the Forum Club, Cape Town, June 1954. Participants were Jordaan, Ngwenya, Simons and Forman. Treason Trial Documents, CAMP Microfilm 405, Reel 20.

43. L. Forman, "Nationalisms in South Africa," *Viewpoints and Discussions*.

44. Ibid.

45. Gail Gerhart and Thomas Karis, *From Protest to Challenge*, Vol. 4, *Political Profiles, 1882-1964* (Stanford: Hoover Institution Press, 1964), 29-30. Forman spent

two years at the International Union of Students headquarters in Warsaw (1951-1953) before returning to South Africa and becoming an editor for *Advance/New Age*.

46. Hirson, "Jordaan," 27.

47. For a discussion of Forman's paper and related criticism in the context of questions of language, culture, and resistance, see Baruch Hirson, "Language in Control and Resistance in South Africa," in *African Affairs: Journal of the Royal Africa Society*, Vol. 80, No. 319, April 1981, 219-238.

48. Jack Simons, "Nationalism in South Africa," *Viewpoints and Perspectives*.

49. Ibid.

50. K.A. Jordaan, "The National Question in South Africa," *Viewpoints and Perspectives*.

51. Hirson, "Jordaan," 27-28.

52. E. Roux, *Viewpoints and Perspectives*, Vol. 1, No. 1, March 1953, Treason Trial Documents, CAMP Microfilm 405, Reel 7.

53. M. Harmel, *Viewpoints and Perspectives*, Vol. 1, No. 3, February 1954.

54. Congress of Democrats, correspondence, "C.T. 15," n.d., COD Microfilm, CAMP Microfilm 1671, Section One, Minutes and Correspondence.

7

White "Democrats" and the Question of Identity

The politics of the national liberation movement in the 1950s had as much to do with culture as with power. Loose circles of activists, journalists, writers, intellectuals and professionals, black and white, liberal and radical, came together socially, sharing a common distaste for apartheid and a common desire to challenge it. Out of these motivations a counter-society developed of non-racial parties and inter-racial romances. This hybrid represented a kind of "no-man's land," what Nadine Gordimer has described as a "frontier." The frontier is characterized by an absence of rules and an atmosphere charged with electricity, undoubtedly a result of the illegality of its proceedings and its contravention of all white South African moral strictures. In a society that was building racial and ethnic barriers at breakneck speed, the need to liberate social discourse and create a non-racial space was all the more urgent, and it was lived as urgent.

What form would the relationships between blacks and whites take in the organizations of national liberation themselves? Nowhere was this question more sensitive and fraught with potential misunderstanding than in the relationships of members of the African National Congress (ANC) and the Congress of Democrats (COD). Blacks and whites worked closely and shared the dangers, defeats, and triumphs of Congress work in these structured organizational relationships. Yet the sub-text of these relationships, the struggle to define one's relationship on this new "frontier" was as important as the manifestations of unity between blacks and whites on the plane of everyday organizational activity. Both blacks and whites, entering the Congress Alliance, were challenging within themselves deeply rooted ideas about themselves and the "other," black or white. The breaking down of organizational barriers and the combining of blacks and whites in common cause against apartheid was, of course, a challenge to all concerned. There were mutual suspicions, antagonisms, feelings of inferiority and superiority and guilt that complicated organizational work.

The thread running through these interactions was that of the individual questioning the very nature of his or her identity. And this was particularly so for white radicals, who, by repudiating "their culture," were faced with the task of redefining a new culture for themselves, of creating a sense of belonging, of identifying with some collectivity or cause that was larger than themselves. Blacks in the multi-racial Congress, who accepted the value of multi- or non-racialism in and for itself, were in a similar position. They were faced with the challenge of stepping outside their generally held understandings of whites, mediating the contradictions they found in their white "comrades," and living with the disappointments engendered as a result of white insensitivity which was itself born of ignorance and the corruptions and distortions of living in an apartheid society. The contradictions and misunderstandings were not a one way street. Blacks could manipulate the guilt and angst manifest in their white counterparts and sometimes did so, out of anger or frustration or pettiness.

The Congress Alliance participants may have viewed the Alliance as an enormously rich and complex socio-political experiment. Because it was an experiment lived in the real world and was thus non-reproducible, the individuals and organizations involved may have felt they had only one chance to complete the experiment successfully. Yet those involved were handicapped by having no blueprint for an outcome. What would constitute success? The realization of government power in the hands of the African majority? The creation of a new society with completely transformed values, where race and ethnicity were no longer aspects of societal discourse? As these questions were contested, without resolution, throughout the 1950s, the relationships of blacks and whites were also constantly open to conflict and change.

Black Congress members and white Congress members were not necessarily thinking about the Congress experiment with the same hierarchy of desired outcomes and changes in mind. For Africans, the national struggle and the winning of power was the primary goal of the national liberation movement. While their white counterparts shared this aspiration, radical whites tended also to focus on the creation of a society characterized by a redefinition of the "nation."

White radicals had much at stake in such a redefinition because the society in which they lived repudiated them completely, just as they had repudiated it. Their hope for finding a place in a new South Africa thus hinged upon redefining the nation so that even while the majority population was in the driver's seat, the nation would be absolutely inclusive, and would return to white democrats full citizenship and a sense of belonging. The socializing that occurred on the frontier had much, in white minds, to do with creating that sense of belonging.

Nadine Gordimer's *A World of Strangers*

Nadine Gordimer chronicles, from within the moment itself, the complexity that characterized the "frontier" between the cloistered worlds of whites and blacks in her 1958 novel, *A World of Strangers*. Gordimer, publishing her first novels in the 1950s, was at the time both a fledgling novelist and a participant in the world of illicit socializing across lines of color and class. Through the African writers at *Drum*, a magazine for Africans publishing work by Africans, Gordimer became a familiar of the Johannesburg literary scene, and was politicized by others involved in it.[1]

A World of Strangers relates Gordimer's perceptions of what was being created, or destroyed, by those involved in the socio-political whirlwind that was Johannesburg in the 1950s, with its small, non-racial black-and-white space hovering somewhere between the segregated plains. The story concerns a young Englishman, Toby, who comes temporarily to take over the Johannesburg office of the British publisher Aden Parrot. Toby makes the acquaintance of a young political radical, Anna Louw, and through her Toby is introduced to the world of socializing across lines of color.

Taken to a non-racial party, Toby makes the acquaintance of an African, Steven Sithole, an iridescent spirit who is loved by all but who has few roots and feels few responsibilities to others. Steven and Toby find a concord immediately: Both want to live private lives. Steven seeks to forget he is an African, his fate inexorably tied to that of the African people as a whole. Toby, brought up in a progressive and enlightened environment in London, is not beholden to that legacy, wanting only to live his life and define his individuality through his actions in the moment. Steven takes Toby under his tutelage, and exposes him to township life, where they find a kinship in drink and forgetfulness. Throughout, Toby carries on a relationship with Cecil, a self-absorbed Englishwoman who draws him into the social life of the British mining aristocracy, with its horseback rides and poolside lunches. Toby is never able to bridge these two worlds. In the end Steven is killed in a car accident while escaping a police raid on a shabeen, an illegal African brewing and drinking establishment. Toby comes to the ultimate realization that it is impossible to remain uncommitted and outside the fray that is the reality of life in South Africa. At the end of the novel Toby makes a personal commitment to an individual, Sam Mofokenzazi, a friend of both Toby and Steve. It is a commitment, at the most basic human level, to see each other through the rigors of times to come. Nothing is guaranteed, but a connection has been made, and no limits are set on the possibilities of such commitment.

Literary scholar Stephen Clingman has characterized the novel's primary fault as its simplistic moralizing, the willingness to define human commitment as capable of overcoming "vast social antagonisms," to be "historically transformative."[2] Clingman acknowledges, and it is extremely important to understand, that this sentiment was authentic and widely accepted among Congress members and supporters and other liberal types in the 1950s. A great deal of stock was put in individual agency, and the simple act of blacks and whites coming together in a social situation could be extremely significant emotionally for the participants.

Anna Louw is the only "radical" in the novel. Louw is a deeply committed person, a former member of the Communist Party, a union organizer and legal aid worker. She is divorced, having been formerly married to an Indian. Toby describes Louw's perspective as:

> [that] of the frontier, the black-and-white society between white and black, and I [Toby] was only a visitor there, however much I had made myself at home. Anna was a real frontiersman who had left the known world behind and set up her camp in the wilderness; the skirmishes of that new place were part of the condition of life, for her.[3]

When Anna takes Toby to his first non-racial party, Toby is moved to remark that to meet people of different colors and races at the party, "to have them there because they wished to be there -- did have, even for me, after one month in their country, the quality of the remarkable."[4] It is to Anna that Toby admits his surprise at meeting Steven Sithole, who is likable, hardly the stereotype of the solemn "bespectacled old Congress gentleman."

In a number of situations, Gordimer presents the problem of blacks and whites attempting to talk as each thinks the other would expect rather than directly *to* one another. Toby is approached, at his second Sophiatown party, by an African man who, pointing to the wildly gyrating dancing couples, suggests that Toby must find such exuberance "crude." If that is what the African would expect any white man to feel, then his reading is incorrect, for Toby feels much the opposite:

> I understood, for the first time, the fear, the sense of loss there can be under a white skin. I suppose it was the point of no return for me, as it is for so many others; from there, you either hate what you have not got, or are fascinated by it. For myself, I was drawn to the light of a fire at which I had never been warmed, a feast to which I had not been invited.[5]

The only space in which neither black nor white enjoyed a particular hegemony was that of the frontier, that black-and-white space that materialized wherever blacks and whites in roughly equivalent numbers came together. Who were the inhabitants of this ephemeral space? To Gordimer's eye they were an eclectic lot. As Toby described it, "there was not room to seek your own kind in no man's land; the space of a few rooms between the black encampment and the white."[6] There were the truly committed, like Anna Louw. There were also liberals who sought to express solidarity and find acceptance in these gatherings, and other whites who socialized with Africans because it was chic, what Toby described as having a "pet" African friend whose name one could throw around at parties.[7]

The challenge was that of overcoming the sense of "otherness," of connecting fully, unselfconsciously, and as equal human beings. This was a very difficult, possibly impossible project, and the inherent difficulties were reflected in the tensions, misunderstandings, and petty slights that came between characters. Toby described the discomfort created by the hostess at one non-racial Johannesburg party. Only with alcohol did the party get into swing, and then the hostess, drunk, leaned into a conversation about the possibilities of a South African *lingua franca* to remark that: "It won't be the whites who'll decide what language is going to be spoken here, it'll be you fellows."[8] The acceptance of the hostess's hospitality becomes, for her, an acceptance of something greater, a personal triumph of bridge-building and an affirmation of personal non-culpability. Yet the hostess, as a personification of many whites on the frontier, had yet to come to grips with the paradoxes of that participation.

Neither the hostess nor Steven were acting as equals. Toby understood the hostess's actions. They were the responses of someone with absolutely no understanding of the "other," and therefore without the ability to define differences and similarities. Both blacks and whites were taking something from these encounters, but it was difficult to define what it was that was taken away.

The novel's core contradiction is thus its abiding belief in the transformational qualities of human relationships even when those who are attempting to meet, to understand, and to transcend keep falling back on conventions or distrusts or frustrations that separate them. After a play, Steven and a number of the other playgoers return to Toby's apartment for drinks. There, Steven taunts Toby for having no apparent interest in African women. Toby reacts, thinking:

> I understood that he meant what he said, it was a cover for some reservation he had about me, some vague resentment at the fact that I had not been attracted to any African woman. He, I knew, did not suspect me of any trace of colour-prejudice; he attributed my lack of response to something far more wounding, because valid in the world outside colour -- he believed that African women simply were not my physical concomitants. It was a slight to him.[9]

Here, in this aggravation, this slight, Toby and Steven nearly reach a point at which individual human emotion takes over from the more common tendency to screen all interaction through racial glasses. The incident above is ambiguous, for Toby understands it as the outcome of a personal slight rather than a racial slight. Yet could it be completely divorced from that? Toby repudiates Steven's standards of physical beauty, and while this could be explained away as personal preference it is also an ambiguity, possibly understood as a stumbling block to "connecting" in the most elemental of senses: the physical. The evening ends when the landlady, who has seen blacks come into the apartment, ridicules Toby's guests and threatens Toby with eviction in front of them.

Shortly thereafter, Toby, returning from a hunting trip, is informed of Steven's death. Prior to this moment there are intimations that Toby has come to understand the impossibility of moving between the black and white worlds while remaining aloof from both: "The only way to do that was to do what Anna Louw had done -- make for the frontier between the two, that hard and lonely place as yet sparsely populated."[10]

Toby is presented with two choices. He is offered the opportunity to stay in South Africa for another year or two. And if he stays, he faces choosing between the two worlds he has experienced. Toby has begun to spend a great deal of time with Sam Mofokenzazi and his wife, meeting "a different sort of people," Congress leaders, young African doctors and lawyers. At the end of the novel, Toby and Sam stand on a train platform. Toby is about to leave for Cape Town on business. Sam alludes to his desire that Toby be his expected child's godfather, and Toby promises to come back in time for the christening:

> I said, ignoring the irritated eddy of the people whose way we were deflecting, "Sam, I'll be back for the baby's christening. If it's born while I'm away, you let me know, and I'll come back in time."
>
> He looked at me as if he had forgiven me, already, for something I did not even know I would commit. "Who knows," he shouted, hitching up his hold on the case, as people pushed between us, "Who knows with you people, Toby, man? Maybe you won't come back at all. Something will keep you away."[11]

Toby appears to make his choice, withdrawing from England and from the South African sub-world of England-in-microcosm with its pool parties and mindless chatter. But the simple act of making that choice is only a beginning, a suggestion. Toby now has to prove himself, as one enjoying the choice of coming or going, seeing or not being seen, involving or not involving. Sam's reaction to Toby's assurances suggests that the expectations raised by this friendship will not be fulfilled, that the lasting connection will not be made. Clingman describes Sam's child, Toby's godchild, as "the fertility of the future in a truly integrated society based on respect and human commitment; likewise it represents a cultural synthesis."[12] Anna Louw, talking to Toby about her marriage to Hassam Bhayat, had said that it was good in the end that they did not have children, for "you can't measure an historical process against the life of the child."[13] Anna is suggesting that any child of a mixed marriage would be sacrificed in the larger scheme of waiting to realize heaven on earth, the end of the race madness. Gordimer's message seems hopeful in the end, for Sam and Toby are, in some sense, doing what Anna and her husband did not do. They are pledging to take mutual responsibility to live up to each other's trust, to bring this child into the world as the product of a common vision yet to be realized in South African society. If Gordimer's ending, and the vision that informs it, appear naïve now, it is a kind of naïveté shared by many who were actively involved in the national liberation organizations or who moved on the perimeter about them.

Living Non-Racialism in the 1950s

Nadine Gordimer's rendering of the "world of strangers" is reflected in the accounts, oral and written, of individuals involved in the events of this period. All accounts of the 1950s, with people's attempts to reach across the breach of race and class and national feeling, evoke both the excitement of connection and the feelings of anger, frustration, and guilt engendered by the structural realities of the society, with its enforced ignorance and self-consciousness. These attempts at connection were an integral part of the Congress Alliance experience. The Alliance, while a political amalgamation, was the site of a multi-racial experiment in a holistic sense. Congress members tested each other's trust, understanding, and sense of responsibility on every level of human interaction. The Congress Alliance was thus much more than the sum of its parts, four political groups organized on a national basis. The Alliance was a site of internal struggle where, in microcosm, all South Africa's peoples attempted to come to grips with the entire edifice of law, sanction, and custom that had kept them apart.

The creation of the Congress of Democrats was in itself an ambiguous victory for the African National Congress. Joe Matthews, a leading ANC figure in the 1950s, stated in retrospect that COD had not played the role envisioned for it by the ANC leadership. Rather than acting as a kind of loosely affiliated auxiliary organization of whites and incorporating a relatively broad spectrum of favorable white opinion, COD's members were basically ANC members; at least, many considered themselves members. If this made COD members outcasts from white society, then COD would be unable to provide the entrée the ANC sought to the liberal community.[14] The ANC probably could not have wanted for or expected a more loyal sister organization of white allies. Yet at times this must have seemed to the ANC like a case of identification in excess. The ANC had hoped for a white organization of sufficient ideological breadth to act as a bridge between itself and both the radical and liberal white political communities. The organization that emerged was decidedly radical.

Beyond political ramifications, the psychological dimensions of this outcome must be understood. The individuals who comprised COD's membership believed in the equality of all South Africans without regard to color. And they lived this belief, taking every opportunity to demonstrate their solidarity with blacks. Therefore, it was hardly surprising that they immediately rallied to the call of the African and Indian Congresses, eager to express in action their support for the national liberation movement. Liberals, whose commitment tended to be much less spontaneous and much more equivocal, refused to partake in an organization that they expected would be led, and ultimately controlled for their own purposes, by radicals. Thus the ANC, while welcoming COD wholeheartedly, felt that of all possible outcomes, the organization that formed was not the most useful from a tactical point of view. While the ANC could not question the loyalty and energy of its white counterpart, the tendency of radical whites to identify so closely with the ANC, and to take up leadership roles so naturally in the working of the Congress Alliance, gave the ANC pause and cause for concern.

Ben Turok, a leader of the Cape Congress of Democrats, noted that he spent more time on committees with members of the other Congresses than he spent on COD committees or working within the white community.[15] Turok and other COD members spent time in the townships as well. In 1954 Turok and ANC women's leader Dora Tamana led a deputation of African women to see the superintendent of Langa location outside Cape Town.[16] Bernard Gosschalk, another COD member in Cape Town, spoke at joint Congress rallies in the townships and locations around Cape Town. Gosschalk was a forceful speaker but not without a sense of humor, as suggested by the extemporaneous talks he gave on Langa street corners:

> Dr. Verwoerd had the cheek to say that because one million Africans are going to be removed under the removal scheme from the urban areas, therefore their lives are going to be made happier. So if any of you want to be happy in South Africa, ask the Government to come and pull down your homes and remove you. This is Dr. Verwoerd's English after sixteen years of study.
>
> All that Dr. Eiselen wants is to leave the Coloured people undisturbed. As a guardian of the Coloured people, he says, the presence of the native in Cape Town leads to moral disintegration of the Coloured people. So if any of you Coloureds are standing next to an African you better move away or else you will be morally disintegrated.
>
> Now one thing about the Government removal scheme, there's no need to catch a train anywhere. If you wait long enough in each area they will send an army lorry to pick you up and take you somewhere else.[17]

What did Africans think about the presence of whites sharing township platforms? David Mgungunyeka, an ANC member, was at a township meeting in 1954 addressed by members of a number of the Congress affiliates. Mgungunyeka noted of that particular meeting on the Cape Flats that "there are all colours here -- blacks, whites, Africans. And it is a good thing when people cooperate. That is our ultimate object that there should be friendship and peace in South Africa." For those Africans who had accepted the Congresses' multi-racial form of organization, the presence of whites in the townships was an affirmation that the struggle was a common one, and the future could also be lived in common. Undoubtedly Mgungunyeka's opinion was not necessarily generalized among township dwellers, and some of Mgungunyeka's neighbors may have been confused or incited to anger by the presence of COD members. Wolf Kodesh spent a great deal of time in the townships and camps of the Western Cape, particularly Elsie's River and Langa. Kodesh was a "foot soldier" rather than a leader, and, given his own background growing up among black South Africans, he found that he had a useful role to play moving between the white and black worlds of the Cape, sometimes spending nights in the latter. Kodesh believed that his very presence was useful in opening lines of trust and greater understanding between people who shared a will to change the country:

> If whites showed they could get in to these townships, and you were willing to, then it enhanced the image of whites, their trust in whites. Because look at the whole media, the whole of life. Blacks were inferior, whites were superior. How do you really bridge it? By playing cricket? You bridge it by doing, and going amongst them.[18]

These contacts were not unilateral. Albert Lutuli, President of the ANC in the latter 1950s, winner of the Nobel Peace Prize in 1960, toured the major cities in 1959 on behalf of COD, speaking to whites. Crowds of over 300 came to hear him in both Johannesburg and Cape Town. Amy Reitstein, then a member of COD in Cape Town, described Lutuli's influence upon his white audience in Cape Town as so great that "Afterwards people seemed to want to touch him."[19] Lutuli's talk in Johannesburg was taped, transcribed, and printed in the booklet, "Freedom is the Apex: Chief A. Lutuli Speaks to White South Africans." Lutuli's success in speaking to white South Africans outside the pale of Congress may have been a result of his ability to reach them using images and contexts that they considered their own:

> To me democracy is such a lovely thing, that one can hardly hope to keep it away from other people. Could anyone really successfully shield off beauty. We don't live in Parktown, but we appreciate the beauties of Parktown. We do. And as we move around Parktown from the townships we pause and admire the beauty. I do. I am not a Johannesburg man, but I pause to see the fine gardens, the beautiful houses and surroundings. I stop and admire beauty.

Can you everlastingly cut off a human being from beauty? Can you ward others off? Can you really successfully do it? I suggest that democracy, being the fine thing it is, the apex of human achievement, cannot be successfully kept from the attainment of other men. I say not.[20]

While Chief Lutuli attempted to speak to willing whites in an idiom that they might understand, that of the cloistered beauty of Johannesburg's affluent suburbs, a small number of whites breached barriers and entered the black world, the world of the townships. Gordimer's character Toby had a number of real-life counterparts, among them Michael Picardie. Picardie was a sometime member of COD and the Liberal Party, and a playwright. Picardie spent time in the townships for a number of reasons. He attended ANC meetings there, was involved in collaborative theater in Orlando township in the mid-1950s, and traveled the night world of the shebeens with his African friend Pakamiza:

I remember going to parties in Sophiatown with Pakamiza. Pakamiza would take me around the shabeens for example. I remember staying the night in Sophiatown once, I mean, parts of it were already being pulled down because of the Western Areas removals, but we always felt at home in Sophiatown....I mean the blacks that were there weren't necessarily going to feel patronized by you, or grateful to you, because they realized that white liberals were just another kind of white, but at the same time they were pleased that there were whites like Huddleston and us, who supported them. So the basis for a multi-racial South Africa was already there in the '50s.[21]

Is the jump that Picardie makes, from his individual experience in the townships to the generalized belief that a basis existed for a "multi-racial South Africa" a reasonable one? In hindsight, it is not, but it does illuminate the mindset of the time, when those who were participating in multi-racial activities believed that such connections heralded the dawn of a future free of racial stereotyping. Picardie's comment, in the context of the experience of the 1950s, is quite similar to Gordimer's ultimate derivation from the experiences of her characters: the possibility of individual connection as socially and culturally transforming. Lewis Nkosi, a member of the *Drum* staff, took up this theme in his writings on 1950s' Johannesburg culture. Nkosi, writing of the Johannesburg premiere of Todd Matshikiza's musical play, *King Kong*, the story of an African boxer, contrasted that reception with the rather "tepid" reception it received in London:

The resounding welcome accorded the musical at the University [of the Witwatersrand] Great Hall that night was not so much for the jazz opera as a finished artistic product as it was applause for an idea which had been achieved by pooling together resources from both black and white artists in the face of impossible odds. For so long black and white artists had worked in watertight compartments, in complete isolation, with very little contact or

> cross fertilization of ideas. Johannesburg seemed on the verge of creating a new and exciting Bohemia.[22]

The potential "Bohemia" was going to be realized less in the product than in the process itself. The importance of *King Kong*, as Nkosi suggests, was the shared nature both of its production and its consumption in Johannesburg. The very act of being a member of a theater audience composed of all races and ethnicities was a source of hope and possibility. However, this type of sharing was the facade, the surface aspect of a complex connection that came only with time and patience, and was always in danger of being lost somewhere between the barriers of apartheid and limitations of the individuals involved. Socializing, when it involved contact across racial and cultural barriers in South Africa, was suffused with political and cultural implications and possibilities. That black and white might desire to interact on the social plane contravened the first principle of apartheid ideology: that the various national groups, as distinct cultures, sought only to be with their own and had nothing to share as equals. There was no common terrain providing for such contact. Given the taboos and sanctions against such contact, there was a variable quotient of self-consciousness at functions where black and white attempted to relate as equals in a highly unequal world.

Contact took place on a nearly daily basis among some Congress leaders, and from this contact, with its attendant difficulties and dangers, sprang friendships. Jack and Rica Hodgson, long-time activists (Jack in the Communist Party and Springbok Legion, both in COD), entertained many of the ANC's leadership regularly in their home. Hilda Bernstein, a member of the Communist Party and then of COD, described these relationships as fundamentally important personally and to the life of her family. Rusty Bernstein, Hilda's husband and a member of the Communist Party, Springbok Legion, and COD, noted that it was one thing to work together on committees and to create over time an atmosphere of mutual respect and trust:

> But to go beyond that, beyond that sort of isolated committee level, and to get down to a stage where people really began to see each other as people, you had to have the social contact. Now it might have been very self-conscious in the beginning and it probably was. One was always conscious of the fact that one was doing something offbeat and outrageous and I'm sure the blacks were just as self-conscious about it as the whites were.[23]

When Hilda Bernstein said that the relationships arising from the political work were critical in making the political work possible, she was pinpointing the subjective aspect of white radical participation in the struggle for national liberation. Radical whites needed to find affirmation in their black counterparts, affirmation that they, white South Africans, were human beings who could play useful roles in fighting the oppressive system with which they were associated. This affirmation could occur on a number of levels and take various forms: taking part as an equal in a position of Congress leadership; winning the acclaim of a township crowd at a street corner meeting; or "breaking through" to forge real and enduring friendships with fellow Congress people. The latter was perhaps the greatest challenge in connecting, black to white. This was so because forging that real friendship required coming to terms with the meaning of both the manifold differences -- of class and race and culture and circumstance -- and the similarities and deficits of a common humanity. Hilda Bernstein noted that social contact and the creation of friendships was easiest with those blacks who had "lost all sense of inferiority," people such as Moses Kotane, the union organizer and communist, J.B. Marks, and Dan Tloome. This is a telling statement. It suggests that one of the problems faced in forging relationships between black and white was that of coping with relations which, in South Africa generally, were based upon paternalism and deference. Lewis Nkosi, the journalist and author, said of the non-racial parties and events of the 1950s that they led blacks to the realization that whites did not live "wonderful and mysterious lives," nor were they generally more intelligent or culturally or morally rich.[24] If these realizations were the outcome of such contact, then clearly Nkosi infers that blacks brought to these relationships-in-formation the burden of such expectations. A colleague of Bernstein's, a trade unionist, related to Bernstein a conversation the unionist had had with a long-time African friend about relationships between blacks and whites:

> And he said, you know, still to this day, when I'm sitting in a white person's home, or talking to them, I'm going like that inside, indicating that he was trembling. So it was that that had to be overcome on the part of the Africans. As well as any sense of patronage that had to be overcome on the part of the whites.[25]

The self-conscious aspect of these relationships, certainly in the initial stages, led to mistakes that could destroy them. There was a fine line between acting naturally and overcompensating, and both held potential dangers, the former of insensitivity, the latter of artificiality. Paul Joseph, a member of the Indian Congress in the 1950s, noted that relations with white comrades was a topic of conversation among blacks from time to time, and that during these conversations various aspects of white behavior would be remarked upon as signs of insensitivity or ignorance. One Congress member was constantly exasperated by the fact that whenever he attended a social function in a white home, he would be plied with food and drink as if he were otherwise denied access to nourishment. Another black Congressman was galled when a white comrade, newly married, organized two socials, one for Congress Alliance friends and another for family and close friends, who happened to be white.

The Congressman, furious at this perceived slight, marched with a number of others down to the family party and was taken to task for it by Joseph.

As committed whites attempted to move between the worlds of white affluence and black struggle, they experienced contradictions at times, between their visions as Congress supporters and their immediate environment and class interests as whites living comfortable lives in segregated cities. Accompanying a number of friends to the home of a COD member for a swimming party, Paul Joseph and friends were mortified to find that they were served tea by African servants in white jackets, trousers, and gloves. These examples are not offered to suggest that such behavior was the norm among COD members. Members of COD were no more advanced in their understanding and consciousness (or lack thereof) than were members of the ANC or any of the other Congresses. Yet in this "experiment" some individuals had a greater understanding of, sensitivity toward, and empathy with their counterparts in the other Congresses. Joseph posed his experiences with Bram and Molly Fischer as counterpoints to the tea described above. The Fischers set themselves apart because they treated their black colleagues with neither paternalism nor humility. Joseph said of Bram and Molly that with them one was among friends. If there was tea to be made then it would be a collective effort without the aid of servants.

There were misunderstandings and slights and alienations on the frontier. There was an overabundance of self-conscious, and sometimes self-serving, action, of attempts to make political points through personal and sexual behavior. Yet, if the Congresses did not succeed in creating a "new Bohemia," as Nkosi described it, there were unique moments of comradeship, of mutual understanding and collective action, of possibilities for a future that were glimpsed only in passing. The Federation of South African Women provided an inordinate number of such glimpses.

Paul Joseph wrote an article in 1954 describing the founding meeting of the Federation of South African Women (FSAW). His article provides a window on a unique moment when not only black and white, but male and female were largely forgotten and Congress members worked unself consciously together:

> A visit to the kitchen showed a hub of activity. You would find John Motsabi, banned secretary of the Transvaal ANC, and Youth Leaguer Harrison Motlana slicing ham (and too often slipping a morsel into their mouths!). Young Farried [sic] Adams was preparing biscuits and munching some at the same time. Leon [Levy] would be washing lettuce while Norman [Levy] would be preparing fruit. The Moola brothers would be washing cups while Stanley Lollan of SACPO was busy with the tea urns. Shy Solly and Abdullhay of the Indian Youth Congress would be tidying up. The women on no account were to see an untidy kitchen....Occasionally Rica [Hodgson] and Beata [Lipman] of the Congress of Democrats would sniff around and pass favorable comments.[26]

The impetus for the creation of the Federation of South African Women (FSAW) derived from women's experience in the 1952 Defiance campaign, in which they participated by the thousands. A meeting in Port Elizabeth shortly after the end of the Defiance campaign drew women trade union activists, African National Congress Women's League (ANCWL) members, and members of the Congress of Democrats to discuss the formation of a national, non-racial women's organization.[27] The women at this conclave decided to hold a national meeting in Johannesburg and to draw in women from all parts of the country. Two white women activists, Ray Alexander and Hilda Watts, played central roles in organizing that meeting which took place on 17 April 1954. The 150 delegates present adopted a "Women's Charter" and pledged to work for women's equality within the broader context of national liberation.[28]

The Federation was undoubtedly one of the most crucial non-racial spaces created on the frontier in the 1950s, and certainly one of the most programmatically successful. Membership in FSAW was by organizational affiliation rather than on an individual basis, and among the distinct organizations affiliating to the Federation were the ANCWL, COD, both the Transvaal and Natal Indian Congresses, and the South African Coloured People's Organization (SACPO), placing FSAW squarely in the Congress camp.

Helen Joseph was deeply moved and energized by her experiences in FSAW and through the relationships she built there. Joseph found her participation in FSAW to be more fulfilling than her participation in any other organization she had worked in prior or subsequent to FSAW because she "was not a white woman doing things *for* black people but a member of a mixed committee headed by a black woman."[29]

One of FSAW's most outstanding qualities, in Joseph's mind, was its steadfast adherence to non-racialism in principle and in action. The Federation's first president and secretary respectively were Ida Mntwana and Ray Alexander who were replaced in 1956 by Lilian Ngoyi and Helen Joseph. FSAW leadership delegations, including those leading the two major women's marches to Pretoria in 1955 and 1956, always included women from each of the Congress affiliates. Prior to the historic women's protest in August 1956 FSAW contacted the Prime Minister and requested that he meet with a FSAW delegation upon the women's arrival at the Union Buildings. Prime Minister Verwoerd replied that he would do so only if the delegation consisted solely of African women. Joseph notes that the Prime Minister's offer was rejected with indignation.[30]

The staunch non-racialism of FSAW was mirrored in Joseph's personal experience, working closely with black women leaders such as Bertha Mashaba, Ida Mntwana, Rahima Moosa, and Lilian Ngoyi. Joseph was a key participant in organizing the anti-pass protest marches to Pretoria of 2,000 women in October 1955 and 20,000 women in August 1956. This activity afforded her the opportunity to work closely with FSAW colleagues on a daily basis, including frequent work in the townships:

> My office was conveniently located just around the corner from Bertha's so we used to leave together after work, with large packets of fish and chips, to set off for the townships in whatever car I could borrow from my friends. Weekends, too, saw us campaigning together in the most distant areas.
>
> Organizing with Bertha [Mashaba] was fun. Tall, bespectacled, lively, she was not yet married so was free to accompany me. We knew all the back ways for me to slip illegally into black townships, though once I almost knocked down a black policeman in the dark, so close was my car. I had no permit to be where I was and I had a shock when I became aware of this dim figure beside my open window. "You nearly kissed him!" exclaimed Bertha and then proceeded to talk us out of this delicate situation.[31]

FSAW's activities afforded South African women the opportunity to build the organization together, to share histories and hardships, and to ameliorate distrusts and misunderstandings. The wholeheartedly accepting atmosphere within FSAW allowed Joseph to feel herself an equal member of the organization.

The atmosphere also guaranteed that Joseph's contributions to discussion and debate would be criticized as openly and extensively as those of any member. The Federation's fostering of such an open forum made it an important learning environment for all participants, as on the occasion when Joseph found her two proposed contributions to the Freedom Charter immediately challenged at a meeting discussing women's demands for the Charter:

> At the Federation Conference we discussed the suggested demands very carefully and only two were dropped. One was the section [put forward by Helen Joseph] calling for better conditions in the "reserves," the parts of South Africa set aside for occupation by Africans, the 13 percent of the land for 85 per cent of the people. I was still ignorant of much that mattered to the African people and had not appreciated that the demand would be for a just redistribution of land, not better reserves. I had accepted, as I accepted so much else, the factual existence of the reserves and demanded, therefore, amelioration of what ought not to be.
>
> The other demand which was rejected was also my contribution -- for better birth control clinics. This was my social work approach and drew lively protest from both men and women. (There were always a few men at our women's conferences, probably out of curiosity.) No one must tamper with the right to bear children, no matter what the social or health consequences. I know that especially in urban areas, health education has brought a somewhat different approach now to birth control, but at that time there were strong

political overtones, a suspicion that the "system" sought to reduce the numbers of African people, while encouraging an increase in the white birthrate.[32]

In fact, Joseph's two contributions were the only two demands dropped from consideration at the meeting. This could be taken as a pointed example of white ignorance and/or insensitivity to African sentiments. Yet the atmosphere within FSAW was such that Joseph was not singled out and made an example of, nor was she alienated by this criticism. She went on to play an important role in the Federation's preparations for the Congress of the People and spoke from the platform during the Congress.

The Federation of South African Women may have enjoyed the special non-racial solidarity that it did because of the specific, gendered nature of its membership. The women of the Federation identified points of shared experience and frustration, and of hope and expectation, that bridged barriers of race and class and ethnicity. The Women's Charter of the time reflects the women's understanding of the oppression specific to black women in South Africa.[33]

However, its aims, among them the achievement of equal pay for equal work, of quality education and adequate nourishment for all children, for the provision of basic amenities to all, and for peace, were aims that all the women of FSAW could embrace. The unity achieved in the Federation based on this common ground allowed the organization to speak to its audience with empathy and authority, and to chart a number of the greatest protest successes of the Congress period as a result.

Yet in charting these non-racial interactions, in the Federation of South African Women, at parties, or stumping in the townships, one is drawn back to the moment, at the end of Nadine Gordimer's *A World of Strangers*, when Toby, about to depart on a week's business trip to the Cape Town area, makes his good-bye to Sam Mofokenzazi amidst the chaos of the Johannesburg railway station platform. He assures Sam, whose wife is about to have a child, that he will return to Johannesburg in time for the child's christening, even if that means cutting his business trip short. When Sam, responding to Toby's assurances of his return, says, "Who knows with you people, man?", he is speaking to the fragility of the black-and-white enterprise as a whole. And, in the end, who did know?

A great deal of faith had been put in the Congress Alliance as the herald of a new political, social, and cultural era, the palpable proof that multi- and ultimately non-racialism were the inexorable tides of the future. Yet, less than a decade later, when the African National Congress, the Congress of Democrats, and other Congress Alliance members and affiliated organizations such as the Federation of South African Women had been banned out of existence and South Africa had been "saved" from multi-racialism, who was left to defend against the implications of Sam's aside?

The Complications of White Affluence

The relationships between black and white members of the Congresses were made complex by a variety of factors. The structural ambiguities of an organization of white radicals in an African liberation movement were manifest. Whites struggled to create identities for themselves that they could live with. These identities revolved largely around notions of complete severance from the dominant white society and identification with the black world. Yet, given the realities of daily existence, the majority of white radicals were to be found living in the white suburbs, enjoying the normal trappings and amenities of that life, while leading a kind of double life that found them at one moment by the side of the pool and at the next in Alexandra or Sophiatown or Langa. Radical whites lived ambiguous lives, but lives that, perhaps ironically, provided important opportunities for the organizations of national liberation by virtue of their connections to privileged society.

One of the most important aspects of the affluence and connections of radical whites in South Africa was white access to the suburbs. In the mid-1950s members of COD were involved in attempts to organize domestic workers in the Johannesburg suburbs. The National Consultative Committee (NCC) set up to coordinate the activities of the Congress Alliance affiliates issued a memorandum to the Congresses describing their special tasks in the 1956 anti-pass campaign. COD's task was predominantly that of enlightening the white public as to the evils wrought by the pass system on Africans. Yet there was also a directive to COD members to "endeavour to establish small groups of domestic workers centered around domestics employed in their homes, and to keep these groups in contact with the ANC."[34] A short article in the local chapter events column of *CounterAttack* in May 1956 noted that the COD chapter in Judith's Paarl had held a barbecue, which was a success not because of the attendance of COD members, but because a large number of African domestic workers had attended. Ester Barsel, a COD member in the 1950s, remembered this aspect of COD work as one of which she was particularly proud:

> We used to have Thursday evening meetings of black domestic workers in the area and that was tremendously successful. On a Thursday night we would have about forty women here at our house, and in other houses as well and we would talk to them about their conditions and so on and they would tell us how hard it is for them.[35]

The life of domestic workers was tiring and lonely, involving long months spent away from other family members. The white family with its coterie of servants was a starkly stereotypical image evoking, in both white and black South Africans in different ways, the whole complex of black poverty and white affluence. Yet, in the anomalous context of the white radical, the family servants became links to the huge unorganized world of urban domestic workers. And these were links that the ANC would, given the structural realities it faced in organizing, never be able to forge for itself. There is some irony in the fact that such meetings in the white suburbs attracted more people than did most meetings of COD members. This particular aspect of COD's work highlights the way in which COD members were uniquely positioned to aid in the organization of Congress. White affluence, under the discipline of African nationalism, could prove useful.

Wolf Kodesh helped to organize meeting places for the ANC leadership in the white suburbs in the 1950s and early 1960s. White sympathizers offered homes and apartments for this purpose. Kodesh had arranged for the use of a flat in Berea, a Johannesburg suburb, on an evening in 1961 during the time that Nelson Mandela was living underground. Ten ANC leaders, including Mandela and Walter Sisulu, were walking down the passageway of the block of flats when Kodesh noticed an old couple peering through their door. Kodesh alerted the assembled African leaders and took Mandela with him. Mandela, living underground and hunted by the South African police and special branch, was in his milkman disguise that evening. Mandela, at Kodesh's apartment, refused the use of Kodesh's bed and slept on a cot, waking at five-thirty in the morning to exercise:

> I saw him sort of getting up and he told me he used to run round the townships, you know. I saw him putting on his long johns, vests, and I thought, no way, you're not going to run around here. He says, who's running, who's going out? And he was running on the spot. For about two hours. And he said, you'll be doing it. And he got me doing it. And so I did it with him for an hour afterwards.[36]

Kodesh also recalled a Congress sympathizer, a white man in Johannesburg, who owned a car dealership and provided Kodesh a change of car whenever Kodesh felt it necessary. This sort of access was critical to Congress operations, as government harassment and intimidation intensified. As in the example above, it is clear that COD members were as useful for those things they had access to through others as they were for their own resources. White radicals were critical fund-raisers for the Congress Alliance as a whole. One of Rica Hodgson's principal roles was that of fund-raiser:

> We went to two kinds of people. One were Jewish businessmen, and one were Indians. The Indians, interestingly enough, were the backbone of giving money for the progressive movement in South Africa. It was very hard to collect money from Indians. You would go to a town and you'd go either with Dr. Dadoo or Maulvi Cachalia, or Yusuf Cachalia or some man, or you'd go with Bram Fischer....They would say, yes, we'll agree on a collection, but then

> you had to go to each one shop by shop, house by house by house, and then it would take you two, maybe three days....But they were very kind to us, and took a lot of chances, I mean, living in the platteland.[37]

White radical access to funds and tools made whites the unofficial journalists, printers, and publishers of the national liberation movement. The ANC had no regular publication of its own in the 1950s, and *New Age*, with a non-racial staff (but largely the work of members of the Communist Party since its inception as the *Guardian* in 1937) acted as the unofficial newspaper of the Congress movement. The Congress of Democrats produced dozens of pamphlets and booklets, some of which enjoyed wide circulation. Pamphlets such as "Educating for Ignorance" and "The Threatened People" indicted Bantu Education, forced removals, and other apartheid injustices, and earned a certain cachet for COD among Africans, both in and outside the cities. Abraham Moholi wrote to COD in March 1955 with praise for "Educating for Ignorance." Moholi was a member of the ANC's Youth League (ANCYL) and a regular reader of *New Age*, with a desire to be "educated politically." Moholi wrote that it was "through reading such literature that a true 'democrat' and a peace-loving person will be able to know the evils of this Nationalist government."[38] Another African, Mr. J.J. Thwala, wrote to congratulate COD on the "Educating" pamphlet and inquired as to possible membership in the Congress of Democrats. Yetta Barenblatt, COD secretary in Johannesburg, wrote to Mr. Thwala that COD was attempting to win "Europeans" to democracy, and suggested that Mr. Thwala should join the ANC if he were not already a member: "Of course there is no need to join COD, because we are working for the same thing in our own organization."[39]

COD was working for "the same thing" but with a panoply of resources that the ANC itself did not enjoy. White radical affluence was truly a double-edged sword, for it was of critical usefulness to the ANC and its Alliance partners. Yet it was also the most visible reminder of the gulf, material and psychological, that stood between black activists, workers, and intellectuals, and their white counterparts.

Perhaps one of the greatest ironies in the 1950s was the state's unwitting role in promoting and, one might even suggest, cementing relationships among members of the Congress Alliance. More than one Congress member believes that the Treason Trial was the single most important factor in solidifying relationships between members of the various Congresses and so creating an Alliance in reality that may have existed more in name prior to the Trial. While the fact of the Trial itself promoted solidarity, perhaps even more important was the combination of the government's seating arrangements at trial and the long, relatively boring nature of the proceedings taken as a whole. The government chose not to segregate the 156 defendants, rather seating them in alphabetical order. Communists sat next to Christians, Indian leaders next to

Coloured foot soldiers, African nationalists next to whites. There was much time to talk, to get beyond stereotypes and find individual commonalities, to explore differences of ideology and culture, and to find ways of living with those differences while recognizing common grievances and shared hopes in the context of a common humanity. The defendants sat together, shared meals together, traveled back and forth to court together, and became a much stronger Alliance for it, at this microcosmic level if not generally. Little did the state know it had abetted the creation and maintenance of a place on the "frontier" within the confines of the state court.

Notes

1. Stephen R. Clingman, *The Novels of Nadine Gordimer: History from the Inside* (Johannesburg: Ravan Press, 1986), 50-51. *Drum* was a popular magazine for Africans published in Johannesburg, and on its staff at one time or another were a number of the area's leading African writers. Anthony Sampson, an Englishman who came out to South Africa to edit *Drum* from 1951 to 1955 was also a friend of Gordimer's. Gordimer also entered a long-term friendship with Bettie du Toit, a former member of the Communist Party, union organizer, and Afrikaner renegade. All of these relationships can be seen as part of Gordimer's "education."

2. Clingman, *The Novels of Nadine Gordimer*, see Chapter 2, "Social Commitment: *A World of Strangers*," 45-71.

3. Nadine Gordimer, *A World of Strangers* (Harmondsworth: Penguin Books, 1984), 175.

4. Ibid., 84-5.

5. Ibid., 129.

6. Ibid., 168.

7. Ibid., 169.

8. Ibid., 171.

9. Ibid., 215.

10. Ibid., 203.

11. Ibid., 266.

12. Clingman, *The Novels of Nadine Gordimer*, 56-7.

13. Gordimer, *World of Strangers*, 176.

14. Interview, Joe Matthews, Gwendolen Carter Papers, Box 39, File 7, University Archives, Northwestern University.

15. Interview, Ben Turok, London, 27 January 1987.

16. *New Age*, 30 December 1954.

17. Police transcript, Congress of the People meeting, Langa, 11 December 1955, Congress of the People File, Popular History Trust, Harare. As with any official or semi-official government document such as a police or trial transcript, one must be conscious of its origins and the manner in which it was created (police transcripts were often written down by an officer on the spot who would attempt a verbatim transcription, the officer's command of the language often being imperfect). The quotations from transcripts here are utilized to provide a sense of the flavor and content of discussion,

and not as material fact. Their quotation should not be taken to represent acceptance of them as records of fact.

18. Interview, Wolf Kodesh, London, 12 February 1987.

19. Amy Reitstein, notes on interview conducted by Gwendolen Carter, 20 January 1964, Cape Town, Box 39, Gwendolen Carter Papers, University Archives, Northwestern University.

20. Albert Lutuli, "Freedom is the Apex: Chief A. Lutuli Speaks to White South Africans," booklet issued by South African Congress of Democrats, Johannesburg, 1959, 4-5. Colonial World Pamphlet Collection, Institute of Commonwealth Studies.

21. Interview, Michael Picardie, Cardiff, 24 January 1987.

22. Lewis Nkosi, *Home and Exile*, (London: Longman, 1983), 17.

23. Interview, Hilda Bernstein and Rusty Bernstein, Dorstone, 25 January 1987.

24. Nkosi, *Home and Exile*, 22.

25. Interview, Hilda Bernstein, Dorstone, 25 January 1987.

26. *Fighting Talk*, Vol. 10, No. 4, March 1954.

27. Tom Lodge, *Black Politics in South Africa Since 1945* (London: Longman, 1983), 141.

28. Ibid., 142.

29. Helen Joseph, *Side by Side: The Autobiography of Helen Joseph* (London: Zed Books, 1986), 5.

30. Ibid., 18-19.

31. Ibid., 10.

32. Ibid., 44.

33. The Women's Charter can be found in Raymond Suttner and Jeremy Cronin, *30 Years of the Freedom Charter* (Johannesburg: Ravan Press, 1986), 161-166.

34. National Consultative Committee, "Memorandum on Anti-Pass Campaign," Clause 3, "Special Tasks of the SACOD, SACPO, SAIC, SACTU and FSAW."

35. Interview, Ester and Hymie Barsel, conducted by Julie Frederikse, held in the South African History Archive, Johannesburg.

36. Interview, Wolf Kodesh, London, 12 February 1987.

37. Interview, Rica Hodgson, London, 12 February 1987.

38. Correspondence, Abraham Moholi, Katlehong Township, Natalspruit, Transvaal, to Congress of Democrats, CAMP Microfilm 405, Reel 5. Daniel Lolwana in the Cape wrote to congratulate COD on the same pamphlet and noted that "When I read a few portions of this booklet to my fellow Africans in a meeting of the Branch, they were so taken up that some were in tears when they heard of Dr. Verwoerd's aims on their youngsters of which they are proud."

39. Correspondence, Yetta Barenblatt, Secretary, Congress of Democrats to Mr. J.J. Thwala, Moorchiba's Cash Store, P.O. Edenvale, 5 November 1956, Treason Trial Documents, CAMP Microfilm 405, Reel 19.

8

COD, the Congress of the People, and the Freedom Charter

If one were to point to a single, specific Congress Alliance campaign most responsible for building unity within the Alliance and promoting non-racialism as a realizable vision in the South African future, that campaign was the Congress of the People, the core activity of which was the drafting of a Freedom Charter. In June 1955 nearly 3,000 delegates and thousands of observers gathered on a field in the Johannesburg township of Kliptown for the Congress of the People. The Congress of the People was the culmination of an idea born nearly two years earlier when Professor Z.K. Matthews, teaching in the United States, found that he had no real answer when asked what those fighting apartheid sought in a new South Africa. Matthews brought his dilemma back to other ANC members in South Africa.[1] In a meeting with his wife, sons, and a number of other ANC leaders at their home in Alice, the idea of a national convention to ratify a people's manifesto was raised.[2] The idea was adopted, and the Congress itself was held on the 24th and 25th of June, 1955, in Kliptown. The Congress was envisioned as the national convention that should have taken place at the time South Africa was created but did not. Its organizers sought the most representative gathering possible, urban and rural, rich and poor, black and white. In fact, members of the National Party were invited but did not come.

For the whites active in the Congress of Democrats (COD), the Congress of the People was a crucial moment in which to prove themselves as Congress loyalists, and to promote the Alliance among whites. As an Alliance affiliate, COD was given equal representation on the organizing body of the Congress of the People and was also represented on the Freedom Charter drafting committee. While COD became an integral member of the national liberation movement in the mid-1950s, the Liberal Party, which had decided to shun the ANC's offer of representation on the organizing committee of the Congress of the People placed itself on the margins of the critical opposition.

The Congress of the People was to be a different sort of campaign, proactive rather than reactive. The aims of the Congress campaign were primarily: to collect demands from the disenfranchised majority that would then form the basis of a Freedom Charter, enunciating the majority's demands for, and vision of, a free South Africa; to hold a Congress of the People, a kind of national convention or alternative parliament at which the Freedom Charter would be discussed, debated, and, hopefully, acclaimed; and to use both the campaign process and its product, the Freedom Charter, to expand the Congress Alliance's base of support. With the end of the Defiance Campaign there had been a notable lapse in organized mass anti-apartheid activity. The campaign against the Western Areas removals was initiated in 1953, and the plan to fight the Bantu Education Act was introduced the following year. Neither campaign was successful, nor did these campaigns create a sense of momentum within either the Congress Alliance or the wider oppressed community.

Because the desires and demands shaping the Congress of the People and the Freedom Charter were applicable to the majority of South Africans, the campaign provided a useful point from which to proselytize and politicize. Thousands of meetings were held in all parts of the country, in cities, towns, and villages. It was the first truly national campaign of the Congress Alliance, and each of the Congresses published its own materials aimed at its particular community, such as this pamphlet excerpted from the Transvaal Indian Youth Congress:

> Listen friend....Yes, it's you I mean -- Amrit, Amina, Marimoothu, or whatever your name is....*Every* young person born in this country is entitled to certain things, no matter who he is or what his race....A call has gone out to the people of South Africa to speak together of freedom....How can such a mighty Congress be organized? Only through a band of voluntary helpers, who will cover the country from end to end, who will enter every town and village, who will go into every factory, every place where people work. These people will be FREEDOM VOLUNTEERS....It is a job, for you, Ahmed, Ismail, or whatever your name is. Don't sit back and dream of what you would like from life. Write your hopes into the Freedom Charter.[3]

The campaign for the Congress of the People, and the campaign following it to collect a million signatures in support of the Freedom Charter, provided myriad opportunities to spread the Congress message and to recruit new Congress members. The Congress of the People demanded a more intensive pattern of contact, mutual planning, and mutual aid than had any joint Congress effort before. Indian Congress member A.S. Chetty related experiences of going into African areas with ANC members to collect demands. The ANC branch in Pietermaritzburg worked closely with the Natal Indian Congress (NIC) there:

> African people were very cooperative with Indian volunteers, in both the urban and rural areas. When we Indians go into an African area there was *no* question, you know, that the guy's going to assault you. The moment you give them the sign, you're a comrade. You say: "Afrika!" and they return it: "Mayibuye!" Straight away you're a comrade. Open, come into the house and talk.[4]

Despite government intimidation and attempts to stop delegates from arriving, the Congress proved representative of the South African population. Of the 2,884 delegates in attendance, 320 were Indians, 230 Coloureds, 112 whites, and 2,222 Africans.[5] The Congress of the People derived much of the legitimacy it enjoyed in later years from the perception that it was largely an organic initiative. Although it was conceptualized and organized by the Congress Alliance, and although the Freedom Charter was actually the product of a small working committee attempting to synthesize the people's demands, the Congress itself was host to a broad range of South Africans in every sense of the word, geographical, cultural, ideological, and economic.

The campaign remained a central aspect of Congress work through the end of the 1950s. The Freedom Charter was a focal point of Congress organization in the latter half of the 1950s and provided a concise statement of common Congress goals and beliefs. In late 1956 the government arrested 156 Congress members who had been involved in the Congress of the People on charges of high treason. The number of accused was later reduced to thirty. The trial ended in March 1961, with acquittal. The Treason Trial engaged much of the Congress Alliance leadership for years, bringing together members of the four Congresses on a national basis and for an extended period of time for the first time in Congress history. The trial taxed the resources of the Alliance and its friends but proved to be an element of continuity in Congress campaigning throughout the latter 1950s. If the Treason Trial was terribly costly in lost leadership and monetary expense, it was also the basis for an enhanced unity among the Alliance affiliates. Not only did the Trial period cement the Alliance, but, as discussed below, it was crucial at a more subjective and elemental level in the process of building a space on the "frontier," breaking down self-consciousness based in color, class, and ethnicity and promoting the enduring personal relationships that stood as symbolic pointers to the possibility of a common future.

Within the ANC itself, the Congress of the People and the Freedom Charter were to prove divisive and largely responsible for precipitating the breakaway of the self-styled Africanist bloc. Africanists understood the Charter as a repudiation both of the ANC's 1949 Programme of Action and of African nationalism generally. Africanists believed that the national liberation movement was diluted when Africans worked with, and shared leadership with, non-Africans. Their antagonism and suspicion applied particularly to the Congress of Democrats. COD was perceived as bearing a double threat: that of white paternalism with its desire for control, coupled with communism whose object was, again, control. Thus, the unity realized in the Congress of the People and enunciated in the Freedom Charter was not unencumbered. The Africanist breakaway in 1958 was just the most visible manifestation of the potential ideological power of a more exclusive African nationalism. The Freedom Charter stood as a monument to the possibilities of unity between the African, Indian, Coloured, and democratic white communities. The Africanists were a reminder that that unity would always be contested.

Members of COD played active roles in every phase of the Congress of the People, including the collection of demands in the countryside and the drafting of the Freedom Charter. The prominence of Coloured, Indian, and white Congress members led some ANC members, wary of cooperation for years, to question whether the Congress of the People campaign was not being hijacked by "foreigners," and more particularly by white communists. Even among solid Congress supporters, there appears to have been antagonism generated in a number of instances by the prominent role taken by whites as explainers of Congress policy.

This wariness of the Africanists and other older, more conservative African leaders was mirrored in the Liberal Party. Founded in 1953, the Liberal Party stood outside the Congress Alliance, a result of the Liberals' anti-communism and unwillingness to accept the immediate goal of an unqualified franchise. The ANC had invited liberals to the 1952 meeting, which led to the formation of COD. The ANC, looking to broaden its base of support among whites, found the liberal community a sensible target for its attentions. With the initiation of the Congress of the People campaign and the formation of a guiding body to oversee the campaign, the ANC grasped an opportunity to bring the Liberals into a working relationship with itself and the other Alliance partners by offering the Liberal Party full and equal representation on the National Action Council of the Congress of the People. The Liberal Party's ultimate refusal to participate in any manner in the Congress was a sign of the tension engendered by COD's presence as a fully vested member of the Congress Alliance.

COD, the Liberal Party, and the Congress of the People

COD's presence on the National Action Council (NAC) of the Congress of the People was a key factor in the Liberal Party's decision not to participate in the Congress. What specifically did the Liberal Party object to in COD's role? Janet Robertson and Douglas Irvine have both suggested that the Liberal Party's ultimate non-participation was a reflection of the party leadership's belief that Liberals were being asked to play a rubber-stamp role in a process whose orchestration was complete.[6] Liberals believed that this *fait accompli* on the part of the Congress organizers was the result of communist infiltration and agitation. Particularly, the Liberal Party was concerned with the prominent role accorded the Congress of Democrats. COD was considered little more than a communist front, and Liberals found evidence for communist control of the Congress at Congress of the People meetings where, they argued, participants were expected simply to express support for Congress plans and listen to discussion of current Alliance programs and peace organization propaganda.[7] In the mid-1950s a substantial number of Liberal Party members believed that the party might yet achieve a popular political role that could be jeopardized if Liberals were to be seen interacting with communists. Thus the Liberal Party leadership, while initially accepting participation on the National Action Council of the Congress, decided ultimately to withdraw completely from the Congress campaign and refused even to send a message of goodwill. The ANC interpreted the Liberal Party's decision as an unwillingness to acknowledge African leadership of the Congress and more generally of the movement for liberation itself. Robertson suggests that the Party's decision estranged it from the ANC, a damaging episode that would show signs of repair only with Liberal aid to the Congress Alliance during the Treason Trial.[8]

This description of the events leading to the Liberal Party's withdrawal from the Congress has been challenged, in some detail, by Dave Everatt.[9] Everatt has argued that while anti-communism was generalized within the Liberal Party and many Liberal Party members dismissed COD members as "cranks," there were members of the Liberal Party who felt strongly that the party must enter the Congress campaign or face marginalization. Everatt suggests that in 1954 and 1955, during the run-up to the Congress of the People, the Liberal Party was in crisis. Party members were struggling with choices: the universal franchise or the qualified franchise; adherence to parliamentary activity or movement into extra-parliamentary activity. The Liberal Party in the Cape was predominantly conservative and sought party adherence to the qualified franchise and parliamentary activity. party members in Natal and particularly in the Transvaal took a more radical stance on the questions of franchise and parliamentary action. Some believed that the party should establish working relationships with COD branches. The virulence of Cape Liberal Party anti-communism is described by Everatt as such that those members considered "radical" (Jock Isacowitz in Johannesburg, Violaine Junod in Natal) by the conservatives were accused of using "communist" tactics. Liberal Party member Jimmy Gibson's election campaign in the Cape, which attracted a number of black adherents to the party, led party chairman Oscar

Wollheim to suggest that Gibson had an ulterior motive in bringing blacks into the party.[10] The Liberal Party was thus in a desperate state of disunity at the time it was offered equal representation on the National Action Council of the Congress of the People.

Everatt, having read extensively in the Liberal Party's correspondence, suggests that the Party Executive Committee's decision not to participate in the Congress, taken in late January 1955, was actually a victory for the party's conservative faction rather than a consensus decision. The conservative victory had come with the passing of a resolution at the Liberals' 1954 National Conference that bound the party strictly to parliamentary forms of action.[11] The Liberal Party did meet with a delegation from the National Action Council (NAC) which included Oliver Tambo, Yusuf Cachalia, and Joe Slovo. Party representatives also attended an NAC meeting in Natal and reported back to the effect that the COD did not control the NAC and that the Liberal Party's presence was very much desired, particularly by the ANC. The conservatives rebuffed this, and Margaret Ballinger, a leading Liberal Party conservative and African representative in Parliament, wrote Z.K. Matthews complaining that the organization and preparation for the Congress were inadequate and it should therefore be postponed.[12] However, in a Liberal Party memo sent to Cape party leaders at the time, the concerns evinced were rather more harshly put:

> Many of the organisers of the Congress are not really interested in the well-being of the people. They are interested in power. They will use any weapon they can find to further their ends. Most of these men are white. They include lawyers who use the grievances of the African people to make their names as "fighters of the people" deliberately fighting hopeless legal cases to establish themselves -- and getting paid handsomely for it. They include men and women who set up bogus organisations as cover for their normal Communist aims....[Their aim is] to rig control of the machinery of the Congress, making it a pure Communist-front organisation.[13]

Everatt does not suggest that some party members were communist sympathizers or held any brief for communism. Rather, there were Liberals who believed that the party's legitimacy in African eyes would rise or fall on the basis of the party's decision to participate in or reject the Congress. These Liberals wanted to fight the communists from within the Congress while using Liberal participation to win new black adherents to the Liberal Party.[14] There were crucial differences within the Liberal Party on the question of constituting the party's base in the years ahead. Party conservatives looked to a white base and the continuation of parliamentary activity. Party "radicals" looked to the African community and by definition accepted the necessity of engaging in extra-parliamentary activity. Everatt sees the ultimate issue as one of control: The conservatives feared that the Liberal Party had nothing to gain in the

Congress of the People, and, lacking the ability to control the process and the product, could be dragged into support of ideological pronouncements and extra-parliamentary actions in which it took no interest. Thus, the Liberal leadership responded to the NAC's invitation by stating that the party would be willing to participate in a Congress based upon organizational representation, believing that the NAC would never agree to this. The Liberal Party leadership was thus surprised to receive a reply from the African National Congress acknowledging the form of representation as organizational. When the Transvaal division of the party received an NAC request two weeks prior to the Congress that a message and observers be sent, Jack Unterhalter, a Transvaal leader, was informed by the Liberal Party's Executive Committee that there was to be no participation of any kind. Neither Transvaal nor Natal Liberal Party constituencies had been consulted in this decision. The Transvaal Liberal Party considered disobeying the EC order but in the end did not do so.[15] Everatt's account does not wholly contradict earlier understandings of Liberal motivations in the 1954-1955 period. Rather, the strains within the Liberal Party clearly suggest that if the Liberal Party membership as a whole was anti-communist, and even if the membership generally believed that COD was manipulating the Congress of the People, there was no consensus as to the Liberal Party's role, or non-role, in the Congress. COD was thus a major factor, if not the sole factor, in the conservative-led decision against participation in the Congress of the People.

COD and the Run-Up to the Congress of the People

Karis, Carter, and Gerhart have suggested that neither COD nor the South African Indian Congress (SAIC) had much enthusiasm for the idea of the Congress of the People when it was first proposed.[16] If this was true of COD initially it was certainly no longer true by the time the campaign began in earnest in mid-1954. An internal COD document, discussing the Congress of the People and the role of COD therein, found important possibilities in the Congress, which included increasing the ranks of the Congress Alliance and of "raising the level of political understanding of the people sufficient to turn the tide of the battle."[17] This rather over-optimistic language was characteristic of COD at the time and suggested that any reservations expressed by COD's leadership had been jettisoned. In a COD discussion paper, there was much talk of COD's particular role, that of "*ensuring the maximum possible European participation in the Congress of the People.*" The COD document suggested that regions and branches focus on local white groups and organizations. COD branches were to write to local organizations and explain the purpose of the Congress of the People and offer to send a speaker. If the offer was accepted, members of the organization in question would then be invited to participate in the Congress and to make a donation. The circular also set out a plan of house-to-house canvassing and street corner meetings. Finally, members were asked to chalk and paste up Congress slogans and write to the press to put the Congress of the People in the minds of their neighbors. In April 1954, a month after the founding conference of the campaign, an unidentified writer asked in a *Fighting Talk* article, "Will the Europeans Be There?" The answer was a resounding "likely." The writer also disagreed with progressive whites who claimed that they were of a different species than their less progressive brethren, who would never change:

> But "we" are different. Like all our fellow South Africans we have been born and reared in an atmosphere of exclusiveness and prejudice. But somewhere along the line our outlooks have been influenced away from the rushing stream of blind race prejudice, perhaps by the ideas of great thinkers; perhaps by the currents of democratic ideas which flourish strongly in lands abroad; and perhaps by the realization, driven home on us by the democratic movement of the non-European peoples, that there is a better way of life for us in South Africa, than that which the Malan government prepares for us.[18]

Yet whites in the Congress movement were absolutely different, and COD branches, which had constituted themselves as Congress of the People local committees were generally unsuccessful in generating interest among whites around the Congress of the People. In a report to the Transvaal Congress of the People committee, a COD representative noted that four branches, Hillbrow, Bellevue, Youth, and Northern had conducted campaign meetings for whites. Hillbrow had conducted six such meetings, Bellevue and Youth had each held two meetings, and Northern had held one, these by late February of 1955:

> The attendance at these meetings, of the people outside the Congress of the People [non-Congress Alliance members] has on the whole been rather small, e.g., 2 meetings called by the Youth Branch, resulted in attendance of 1 person. Bellevue Branch contacted many for one meeting; only three attending. A meeting in Orange Grove called by Hillbrow, had an attendance of 6 people outside of COD and was fairly successful, but a meeting in Hillbrow of 20 people contacted, none attended....We have therefore reached the conclusion that it is very difficult to get Europeans to attend meetings.[19]

Radical whites were generally ineffective among the white population, even among those identified as the most likely potential recruits. While frustrating to COD members, given their mandate to work among the white community, the inability to reach that population with the Congress message fed naturally into the desire of white radicals to work with and among blacks. Time spent proselytizing whites reinforced long-held cynicism as to the difficulty of changing white hearts and minds. Working in the black community, Congress whites felt as if they were integral, equal, and appreciated members of the force for national liberation. Whether working in a township, collecting demands in a rural area, or taking part in an executive meeting, radical whites could claim an identity with the majority and with that which they held to be politically and morally correct. Working door-to-door in white neighborhoods reinforced the crisis of identity, the sense of ambiguity adhering to the individual caught between two worlds. Amy Thornton, a member of COD in the Western Cape, had been going into areas outside Cape Town prior to the Congress of the People campaign, and then began to work in the campaign:

> Over the Congress of the People campaign the different Congresses -- African, Indian, coloured, and the white COD, really pulled together for the first time. In Cape Town we had a Joint Congress Committee. I was secretary of the Cape Town JCC. The Indian Congress didn't really function in Cape Town, the Indian community was so tiny. But three of us, representing the other congresses, would go out to the country districts in the Western Cape. We'd usually have a contact person in the different dorpies. A small meeting would be convened. We'd speak about the Congress of the People, we'd leave leaflets behind, and the local comrades would then organise the collection of demands.[20]

The working unity established among the Congresses in the Cape boded well for the success of the Congress of the People. Still, there were latent tensions surrounding the participation of non-Africans that could be exploited by individuals or organizations who were angered by this multi-racial display.

Tshunungwa in the Cape

Tensions arose from the prominent role played by COD members generally and particularly in African areas. The ANC had appointed T.E. Tshunungwa as national organizer for the Congress of the People, and it was Tshunungwa's job to travel the country, aid in the organization of the campaign, and report back to the National Action Committee. In January of 1955 Tshunungwa traveled to the Western Cape to review the progress of the ANC branches there. Reporting back to the NAC, Tshunungwa complained that the campaign had been poorly explained to the people of the Western Cape. Tshunungwa alluded to "unhealthy [read strongly African nationalist] utterances" from some anti-white Africans, what he described as a kind of "confusion":

> What has caused this extreme form of confusion is to find the COD men taking a lead in the ANC meetings. We are the vanguard in the liberatory struggle and our allies must not be exposed to unnecessary attacks by those of our people who are not yet well educated about this basis of cooperation. A

> politically raw African who has been so much oppressed, exploited and victimized by the European sees red whenever a white face appears.[21]

The tone of Tshunungwa's report was ambiguous. While Tshunungwa appeared to defend COD members against the wrath of "politically raw" Africans, he also clarified, with emphasis, the primacy of the ANC and African nationalism in the national liberation movement. Tshunungwa's cautionary note could have been understood as a warning, to the effect that white radicals did not know their place and were stepping outside the boundaries of acceptable practise. Tshunungwa was highlighting the ambiguity of the white presence, its anomalous nature and the potentially justifiable anxiety among Africans who were faced, once again, with white leadership.

In early April, shortly before the Congress of the People, three members of the Cape Western Action Council, the joint campaign committee of the Congress Alliance in the Western Cape, wrote to the National Action Council in Johannesburg about certain problems attendant upon Tshunungwa's visit. The letter, signed by Reginald September, Ben Turok, and John Mtini, representing the South African Coloured People's Organization (SACPO), COD, and the ANC, respectively, claimed that Tshunungwa "created a serious atmosphere of uncertainty and hampered the work of the Congress of the People."[22] The signatories claimed that at a meeting of the ANC's Cape Western Regional Committee, Tshunungwa had attempted to undermine the Cape Western Action Council. Tshunungwa had claimed that the ANC had been kept from playing its rightful leading role in organizing the Western Cape for the Congress. Tshunungwa had also suggested that the Action Council had been formed without his permission and took its directives from "the wrong source, meaning Johannesburg." Apparently Tshunungwa also accused the chair of the ANC Regional Committee of lying. Clarifying the matter in correspondence, the three Congress affiliate leaders stated that in fact the executive of the Cape Western Action Committee included five ANC members, three COD members, and two SACPO members.[23] The correspondents claimed that Tshunungwa had "tried to damage the unity and the equality in action of the sponsoring organizations." The claims of the correspondents suggest Tshunungwa had understated the breadth of anti-"foreign" feeling when he pinned it only to "politically raw Africans." It was reasonable to expect that some Africans would be antagonistic upon finding a white face taking an apparently leading role at an African meeting. Tshunungwa appears to have been among them.

The experiences of COD members noted above suggest that there had been some durable contact between whites and blacks in the Western Cape. John Mtini, one of the leading members of the study group in which Thornton and Kodesh participated, had been a signatory to the complaining letter. If COD members had overstepped the bounds of acceptable behavior in the eyes of some

ANC supporters while taking an active role in Congress campaigning, it is telling that the three Congress affiliates in the Cape chose to write to the national body to complain of attempts to damage their unity in action. The decision to write a letter to the National Action Council also raises an important problem. Were the Congress affiliates asking, essentially, that the ANC mediate between its white allies and its broader African constituency? If so, the ANC found itself in a delicate and complicated position.

There was a real fear on the part of some ANC members that "foreigners," which could mean any non-African or communist, would attempt to hijack the Congress of the People for their own ends. This sentiment was not unique to the Africanists, whose resentment of the ANC's alliances with whites and other non-Africans would intensify in the years following the Congress of the People and the adoption of the Freedom Charter. There were older non-Africanist ANC members who were also wary of the ANC leadership's proclivity to give responsibility and power to non-Africans. One week prior to the Congress, M.B. Yengwa, a long-time ANC leader, wrote to Dr. W.Z. Conco, another long-time ANC member and official:

> Now I do not know how the organisational set-up of the Congress of the People is at the moment, and whether the ANC will be able to give a lead all the time, and not allow itself to be compromised by other members of the Congress of the People. Again I feel that we should be careful not to allow the Congress of the People to be used as a platform for any other campaign other than of the drafting of the Charter of Freedom. There will be sections that will do much in their power to have the Congress of the People become a permanent body that will coordinate the activities of the various members of the Congress of the People. We are not yet ready for such a body.[24]

It was not the intent of any of the Congress campaign participants that the Congress of the People be used as a springboard to a permanent commissariat directing the activities of the Congresses. Yet Yengwa's sentiments were similar to those of leading Liberal Party conservatives in their fear that "sections" of those involved, a coded reference to communists and, particularly, white communists, would attempt to use the Congress of the People for their own aims.

The prominence of the Congress of Democrats in the Congress of the People campaign has been alluded to by other commentators.[25] Only the boldest antagonist could have denied, fears of white usurpation set aside, that COD members played an integral part in the realization of the Congress of the People. Not only did COD members travel to the rural areas and speak in the townships, but they were in great part responsible for the financing of the Congress. The Transvaal Provincial Committee's Treasurer's Report in February 1955 noted that of £46 collected to date, all of the funds had been collected by COD.[26] In the next two months the general funding situation was to change for the better, and this was reflected in the minutes for the Committee's meeting on the 24th of May: COD, £67:19:11; Transvaal Indian Congress (TIC), £36:3:10; ANC, £22:12:8; and SACPO, £5:0:0.[27]

The Congress of Democrats took its responsibilities to the Congress of the People seriously, as evidenced in its fundraising and its committee participation. COD's ability to raise three times the amount raised by the ANC reflected the relative financial strength of COD members themselves, some of whom undoubtedly contributed pounds or were able to fundraise among affluent supporters. However, more was involved than simple questions of relative affluence. When COD members embraced the Congress of the People idea, they did so wholeheartedly. The Transvaal Provincial Committee provided equal representation to each of the Congress Alliance members if they so desired it, and five seats were allotted to each of the affiliates. Attendance figures are available for ten of these meetings, which took place between the 25th of January and the 24th of May, 1955. Given a possible total of 50 representatives present from each organization over the ten meetings, the total attendance figures for each of the Alliance affiliates were: ANC, 27; Transvaal Indian Congress, 27; SACPO, 9; COD, 35.[28] The prominence of COD members in these and other endeavors, including the committee to draw up a draft charter, and the willingness of COD to speak up and make suggestions on various aspects of campaign organization and content, left COD members open to the charge that they were arrogating too large a role to themselves in the process. In a letter from Tshunungwa to COD in Cape Town prior to the acrimony detailed above, Tshunungwa acknowledged receipt of a letter from COD in the Cape and stated in his letter that "we all appreciated the suggestions in your letter based on grounds of expediency in carrying out the campaign and also to have systematic organization and administration."[29] Apparently COD was not shy about offering its opinions and advice, and Tshunungwa's later tète á tète with the Cape Town committee may have been a reaction, in part, to his perception that COD was overstepping its mandate. The Tshunungwa incident is evidence that the ANC's willingness to give COD free rein in the city and countryside opened the ANC itself to charges that it could not control its allies and might lose control of the movement for liberation.

At the Congress of the People

Members of COD were probably aware of these concerns about their participation and may have been troubled by them. Still, like their counterparts in the other Congresses, they held high hopes for the Congress of the People. At a Congress campaign conference in July 1954, Joe Slovo, a leading member of the South African Communist Party, spoke of the possibilities arising from the Congress campaign. Slovo envisioned the creation of "organized units" throughout the country that would stimulate Congress activity. Slovo described the Freedom Charter as capable of doing away with "the feeling of isolated defeatism which is the result of disconnected and sporadic acts of political struggle."[30] An internal COD political report written shortly before the Congress of the People noted that the campaign had already achieved to a degree varying from district to district, some of its objectives. These included the revival of Congress activities and creation of local committees in many politically moribund areas, and the cementing together and individual revitalization of the Congress Alliance affiliates.[31]

The two days of the Congress brought together nearly 3,000 delegates and an estimated 7,000 observers on an empty dirt lot in Kliptown, fifteen miles outside Johannesburg. A COD delegate described the scene:

> There are faces here that are known throughout the land, and names that are as familiar as old friends. There are trade union leaders and clergymen, doctors and farm labourers, mechanics and school teachers, students and social workers, domestic servants and politicians....They are here to speak for others, for ordinary people of a hundred different slum towns and garden suburbs, of factories and farms and villages, for perhaps half-a-million people who have elected them. As they pass into the delegates enclosure they are handed a draft Freedom Charter. They have not seen it before, but yet it rings familiar....There is a spirit of unity and amity here which could not be found anywhere else in the whole area of South Africa. The delegates meet together, speak together as equals, with respect for each other despite their race and social differences, treating each other simply as citizens. In its way, this is the most truly national convention in all South African history.[32]

The Congress had been envisioned, for the purpose of introducing the Freedom Charter, as a kind of gigantic seminar, in which each clause of the Charter would be discussed, debated, and possibly modified as a result. The realities of time and the immense number of delegates, each with a right to speak, prohibited the realization of such open debate. Rather, each clause of the Charter was read and a small number of delegates who had sent up slips of paper requesting the right to speak on that particular clause were allowed to do so: perhaps three or four at most per clause. Each clause of the Charter was introduced by a member of one of the Congresses, and a number of COD members spoke from the stage. Ronald Press read messages to the Congress, including one from Chou En-Lai; Piet Beyleveld presented the *Isitwalandwe* award to Chief Lutuli's daughter and also introduced the preamble. Ben Turok moved the clause regarding the sharing of the country's wealth. Sonia Bunting moved the section on equal rights, and Leon Levy brought a message from the South African Peace Council.

One of the most interesting communications from the podium was Ezekiel Mphahlele's reading and discussion of the clause on culture and education.

Mphahlele dismissed Bantu Education, saying that he refused to teach his children that Shaka was a murderer or that they were inferior because they were Africans. Perhaps the most striking aspect of Mphahlele's talk was his discussion of nationalism. He faced the problems of cross-cultural enmity directly, asking his Indian friends to tell their friends "who own the Majestic Theatre to go and open this cultural organisation for the black man, for the coloured people, to open it for other non-European friends." He challenged Africans to teach their children "to respect the coloured man, the Indian man," so that a culture could be built based upon a nationalism "in democratic form."[33] Mphahlele, as other speakers before him, did not mince words when it came to the oppressor. He rejected the notion that the oppressed sought equality with the white man, suggesting that it was necessary to be better, for the white man had "made a terrible mess of everything." Mphahlele's rejection was not a rejection of a people as a whole but of a culture which he believed to be devoid of value by virtue of its closed nature, its selfishness, and parochialism. Perhaps, because he was a writer, Mphahlele was drawn to consider issues of culture more deeply than others. In completing his discussion of education and culture, Mphahlele looked beyond the Congress Alliance and beyond, even, considerations of conglomerating the Congress affiliates in a unified organization:

> I am looking forward to a day when our culture will so much unify us, we shall no more talk of the Congress of the People as an organisation of Coloureds, Indians and Africans and Europeans. We shall have one movement. We shall have absolutely no distinction, and we will stand together for a united cause. Our culture -- this culture is now growing up. It will not be a culture of Indians, of the Africans. It will be a culture of the people of South Africa, and it is this culture that is going to grow up now.[34]

Chief Albert Lutuli, President of the ANC but unable to attend because of illness, shared Mphahlele's sense of the importance of the Congress as a stepping stone to a more unified culture. In a June 1956 article Nelson Mandela quoted Chief Lutuli's thoughts as to the significance of the Congress of the People. At the time, Lutuli, who was disabled by a stroke and unable to attend, noted that the Congress would signal a first in South African history, a moment when democrats, without reference to race or ideology or party, would renounce racialism and unite under a common program.[35] Mphahlele and Lutuli were intuiting the possibility of a revolutionary transformation that would create a new nationalism, a broad, colorblind South African nationalism. Rights would not have to be discussed on a national basis because they would be the property of all. There would be a South African culture taking from each of the "national" cultures and creating a hybrid that would be revolutionary in that no group could claim hegemony by virtue of its existence, but in which all were endowed. If this vision was open to charges of naïveté it was nonetheless powerful and would be taken up by others in the decades to follow.

The Congress of the People was a critical moment in the history of the national liberation movement. For some it was a moment in which South Africans came together and spoke of freedom, of the injustices that had been done them, and of the conditions necessary for creating a livable and a just South Africa. Those people were moved by the experience itself, given the mass non-racial and democratic nature of the crowd. Others, those who felt the Charter and the Congress Alliance itself were a betrayal of the ANC's legacy -- as expressed in the ANC's earlier documents, "Africans' Claims" and the "Programme of Action" -- were angered. The Congress of the People and the Freedom Charter provided direction to activists who embraced them. However, there clearly was no consensus as to the direction the national liberation movement should take or the precepts that should guide it. The Freedom Charter would prove a point of contention around which dissent raged in the following years.

The Freedom Charter and Dissent Within the Congress Alliance

The process surrounding the Charter's creation proved highly controversial. The Liberal Party had excluded itself from the whole Congress enterprise, one of the party's greatest fears being that the Charter distilled from the drafting process would be a radical document wholly unrepresentative of Liberal philosophy. The Liberal Party would not be disappointed. The decision-making body of the Congress of the People, the National Action Council, manifested the same egalitarian structure as that of the Congress Alliance. Each Congress was provided equal representation on the Council. By placing the authority to draft the Charter co-equally in the hands of the four Congresses, the ANC leadership was consciously ceding the right to define the nature of the liberation movement. The Charter thus produced by the Congress drafting committee was, by definition, a departure from the ANC Youth League's 1949 Programme of Action, and brought to a head the conflict between the Africanists, advocates of an exclusive African nationalism, and those ANC leaders and members who supported the multi-racial alliance and the non-racial vision incarnate in the Charter.

The Draft Charter

The process of collecting demands for the Charter has been discussed. Volunteers collected thousands of slips of paper from across the whole of South

Africa inscribed with the people's demands. From these a Freedom Charter was to be distilled, and a special committee, composed of members from each of the Congress Alliance affiliates, was created to draft the Charter:

> A drafting commission was set up to take all these thousands of bits of paper and try and make some sense of them....And then, from that, came this draft charter. Anyway, this is what the Charter actually emerged from. And then obviously, this drafting commission, in trying to put the thing together had, A, to put it into intelligible language, and B, had to try and put forward, or to fill, some of the gaps. For instance, if you have fifty demands saying we want more land, or we want land, what does the drafting commission do with all this? I mean, it's no good putting in the Freedom Charter the demand, "we want land," who wants land, who's it for, who's going to get it, where are you going to get it from? Are you going to confiscate the land, are you going to buy the land? See, the commission obviously had to do some thinking of its own....It wouldn't be true to say that the Freedom Charter was just compiled, it was a drafting based on that. And the idea was it would be debated line by line or clause by clause at the Congress, and if anybody disagreed with the formulation, it was going to be a free-for-all to disagree and reformulate, but at least somebody had to put forward a precise formulation which encapsulated these demands, and that's what it was all about. Now, at the Congress of the People itself it didn't actually quite work the way it was intended.[36]

An examination of the preliminary draft of the Charter highlights the difficulty the drafters faced in reflecting the demands of the people while maintaining a certain coherence. Preliminary notes on the draft Charter suggested a structure for the document that would present a "general clause of rights," to be followed by particular demands that would juxtapose vision (for example, "freedom of movement") and reality ("the pass laws must be abolished").[37] This basic structure was in fact maintained in the Charter presented to the Congress of the People. However, much of the draft Charter's language was rough, and marginal notes questioned aspects of the text such as language usage and wording, questions of detail and the inclusion or non-inclusion of certain demands. Clearly, the committee was struggling with many syntactical and substantive, though not necessarily ideological, questions. To provide a sense of the metamorphosis of the Charter from draft to final presentation, it is illustrative to look at one of the Charter's clauses as it appears in the draft Charter and in its final form. In the Charter adopted at Kliptown, the first clause following the Preamble states:

> **The People Shall Govern!** Every man and woman shall have the right to vote for and stand as a candidate for all bodies which make laws; All the people shall be entitled to take part in the administration of the country; The rights of the people shall be the same regardless of race, colour or sex; All bodies of

> minority rule, advisory boards, councils and authorities shall be replaced by democratic organs of self-government.[38]

In the earlier draft Charter, this particular area of concern was also the first clause after the Preamble, but had yet to be honed:

> *We South Africans hold that the will of all the people shall be the basis of the authority of our government.* White men and Black, all South Africans, must join hands as partners and equals for the future. Every adult South African, regardless of colour or sex, of national or social origin, should have the unqualified right to vote for, and to stand for election to, the Parliament of our land, and to all national, provincial, and local governing bodies. Puppet advisory boards, 'Bantu authorities' and tribal councils should be abolished. The bonds between the people, the organs of public opinion and their elected representatives should be close, and consultation between the people and their representatives should prevail at all levels and at all times.[39]

Contrasting the two clause drafts, little had changed and nothing was done to alter the meaning or intent of the original. A marginal note on the draft Charter at this clause asked if more details of political rights were needed. Yet the final draft pared some of the sentences from the first draft, including the general statement, "White men and Black" and the final sentence in the draft Charter clause, "The bonds between the people." Whereas the draft Charter contained twenty-eight clauses or clause headings after the preamble, the final Charter contained only ten. Many of the clauses listed in the draft Charter became themes within the final ten clauses, including areas of concern such as trading rights, the role of the police and army, the treatment of the old and sick, and the right to medical care. Of the twenty-eight clauses proposed in the committee's draft Charter, only seven found no expression, or no near-equivalent expression, in the final document. These seven clauses concerned specific rights for farm tenants, equal access to the civil service, the opening of the professions, no taxation without representation, full and equal rights for women, recognition of the family as the fundamental social unit and so entitled to protection, and the provision of free technical and vocational education to workers. Grounds for the ultimate exclusion of clauses may have included over-specificity, over-generality, or the fear of redundancy. Reference is made in the Charter to the rights of women in a number of areas, including equal pay for equal work, maternity leave, pay for working mothers, and free medical care and hospitalization for all, with special reference to mothers. Still, it is somewhat surprising that a specific section on women's demands was not included, particularly as the Federation of South African Women (FSAW) had adopted a "Women's Charter" at its inaugural conference, and the Federation had compiled a document, "The women's demands for the Freedom Charter," which it submitted to the Congress of the People preparatory committee. Many of the women's demands paralleled general demands and are in that sense included in the final Charter, while other points the women had made, some in their own Charter, relating to laws and customs that defined women as minors, were not addressed in the final Charter.[40]

Marginalia suggest that the committee had divided the demands received into categories and then written them up on various sheets. Comments made reference to the sheets and the need for inclusion of the specific demands on them. A short note on the bottom of the last page of the draft acknowledged that the demand for the right to brew and drink liquor had been omitted, and "others left our [sic] can be seen from the sheets -- some omitted purposely, others should probably still be included." Many of the people's demands reflected the immediacy and urgency of their situations. Women brewed and sold beer as one of the most effective means of squeezing an income from, or supplementing an income within, the township economy. However, the government maintained its own beer halls, and individual brewing was illegal and constantly subject to government raids. The legalization of brewing would have been considered a critical demand by some Africans. Perhaps this particular demand was considered too specific by the committee. Yet the fact that it was mentioned on the draft suggests it was a popular demand and provides some insight into the type of judgments and exclusions the committee was forced to make in drafting a reasonably concise document.

Claims made by those outside the process to the effect that the Charter was written by a cabal are difficult to uphold in light of the knowledge of the process. Comments inscribed on the draft Charter and a comparison of the draft and final charters indicates that the committee members made an effort to include as many of the demands received as possible, while creating a document of manageable size and sufficient generality. The one known exception to this, and the most controversial, was the nationalization clause.

The Nationalization Clause

The delegates to the Congress of the People never received the opportunity envisioned by the Charter's drafters to comment and debate the Charter at any length during the Congress of the People. For the majority of Congress supporters this was of no apparent concern. The Charter was unchallenged at Kliptown, and clause after clause was adopted by unanimous acclamation. However the general economic clause of the Freedom Charter proved a point of great controversy and forced a number of Congress leaders to explain away what some observers considered its more sinister socialist implications:

> **The People Shall Share in the Country's Wealth!** The national wealth of our country, the heritage of all South Africans, shall be restored to the people;

> The mineral wealth beneath the soil, the banks and monopoly industry shall be transferred to the ownership of the people as a whole; All other industries and trades shall be controlled to assist the well-being of the people; All people shall have equal rights to trade where they choose, to manufacture and to enter all trades, crafts and professions.[41]

Known as the nationalization clause, its implied dedication to nationalization of major capitalist production and finance was taken as evidence by some of communist hegemony within the Congress movement and communist control of the Freedom Charter drafting process. The draft Charter's nationalization clause was not as extensive as the final product, although it was similar in intent. Demanding that "all should share in the national wealth," the draft Charter clause stated that "The land, the forests, the mines of coal, gold and diamonds, the quarries, the factories and workshops, the railways, should be worked for the benefit of the people. The mines should be nationalized." It appears that the demand for a limited nationalization was present from the early stages of the committee's work. However, Ben Turok, a member of COD who was also a member of the National Action Council, claims to have reviewed the draft of the Charter the Friday night before the Congress was to begin. Turok, finding the economic clause unsuitable in its then-present form, drafted an amendment, which was seconded by another NAC member and written into the Charter.[42] That amendment now stands as the nationalization element in the third clause of the Charter.

Where the draft Charter had called only for the nationalization of the mines, Turok's amended clause demanded the nationalization of "the mineral wealth beneath the soil, the Banks and monopoly industry." A typed marginal note on the draft Charter, referring to the sentence, "The land, the forests, the mines of coal, gold and diamonds, the quarries, factories and workshops, the railways, should be worked for the benefit of the people," suggested that the phrase "be worked for the benefit of the people" was "too vague." Was this sentiment, pertaining to this particular clause, generalized within the drafting committee? Apparently not, for Turok took it upon himself to offer the amended version at the eleventh hour, shortly before the Congress was to begin.

A white former communist thus claims to have been responsible for introducing into the Charter just the type of ideological element most feared and disliked by those -- white liberals and Africanists -- who were most initially distrustful of the Congress process. Was Turok's alleged action a vindication of those fears? The answer is neither a simple yes or no. Turok introduced the amendment, but it still had to be seconded and amended. Other committee members from other Congresses must have agreed to Turok's amendment. And after the Congress the Charter was returned to the four Congresses and the South African Congress of Trade Unions (SACTU) for amendment and ratification. Was Turok taking advantage of COD's equality on the drafting committee and was he in fact acting for COD in offering the amendment? Undoubtedly the amendment was ideologically in synch with the thinking of many of COD's members. However, COD as a body had taken no position on the question of nationalization prior to the Congress of the People, nor was there reason to believe that all COD members would have agreed with Turok's amendment. Turok's rhetoric in public speaking appearances was often characterized by a strong class-analytical component, which marked him as a communist or sympathizer. A few excerpts from the police transcript of Turok's introduction of the economic clause at the Congress provides a flavor of Turok's rhetorical flourish in this regard. Speaking of the mine owners after liberation, Turok stated that:

> They will not have these lovely big Buicks that they drive around in. The whole system of the big factories and the Gold Mines in this country are the enemies of the people. When you walk down one of the streets in Johannesburg, you see a very impressive looking building, and outside there you see various banks....That money, friends, does not come back to you. It goes to our friends living in Lower Houghton. Let the banks come back to the people, let us have a people's committee to run the banks.[43]

It is quite possible that Turok's eleventh-hour amendment was more an act of personal than of programmatic initiative. Nonetheless, Turok's offering would have fueled the notion that whites or communists or white communists were playing key roles in defining the policies and direction both of the Congress Alliance and of the African National Congress, had Turok been identified as the progenitor of the clause at that time.

Members of the Liberal Party were among the most vociferous critics of the Charter, largely because of the nationalization clause. Jordan Ngubane, a leading member of the Liberal Party, claimed that the Charter's purpose was to "condition the African people for the purpose of accepting communism via the back door."[44] Among the Congress inner circle, Lutuli, still disabled and in Natal, was concerned about the apparent ascendance of leftists in Congress. In a letter to Arthur Letele, a doctor and Congress organizer, Lutuli warned that he would consider resigning the ANC presidency were the ANC to "tie ourselves so fast to the Congress of Democrats" that groups outside the Congress Alliance would be spurned if they happened to disagree with a clause in the Charter.[45] Others suggested that the whole campaign leading up to the adoption of the Charter had been framed by communists in such a way that the Charter could be manipulated and accepted as a *fait accompli*. In the wake of this controversy members of the Congresses worked actively to dispel the belief that the Charter was a socialist document. Speakers' notes provided to COD members by the organization described the nationalization clauses as applying to a number of liberal-democratic states and noted that state-run industries were to be found in non-socialist nations, including South Africa. Michael Harmel, a leading member of the Communist Party and a member of COD in 1956, discussed the Charter and the nationalization clause in an article for *Liberation*:

> When our Freedom Charter speaks of the need to transfer monopoly-owned resources into public property, it is not because the framers and supporters of the Charter are all Socialists (they are not by any means), but simply because all the other provisions of the Charter, looked at from any realistic point-of-view, are bound to be ineffective, illusory and unthinkable, so long as the keys to the country's economy remain in the hands of the present gold and land monopolists, who, in their hunger for cheap labour and through their commanding influence, are responsible for all the ills which beset our land and which the Charter seeks to cure.[46]

Michael Harmel argued that one could not speak realistically of the majority of changes envisioned in the Charter without recourse to some degree of nationalization. If this was a reasonable argument to make in the context of an apartheid economy, it did little to lessen the controversy surrounding the clause. The nationalization clause would put the Charter's supporters in a defensive role for years to come. If it had been made public at the time that this clause had been largely the work of a member of COD, this would have been taken as proof by many Congress detractors, and probably a number of supporters also, that white leftists had brought a hidden agenda to the highest levels of Congress discourse and policy-making. Nelson Mandela, speaking from the dock at the Rivonia trial in 1964, said of the clause that nationalization would take place in the context of a private enterprise economy.[47] Nothing in the Charter suggests that its framers were seeking to introduce a socialist system while doing away with capitalism. Jack Simons, a prominent member of the Communist Party, said of the Charter in 1961 that it was "a statement of principles for a freely competitive capitalist society."[48] The controversy surrounding this clause of the Charter had as much to do with tensions within the Alliance and between the Alliance and other organizations as it did with the clause itself.

Critiquing the Charter: Africanists and Other Skeptics

In a form letter from one of the Congress of the People local committees to Africans in rural areas prior to the Congress of the People, the Freedom Charter was described as a "Bible" to be compiled for the struggle, which would act as "the last will and testament to Africans and all the races living in South Africa now and hereafter."[49] However, this Freedom Charter would not be accepted as the holy writ of struggle by all ANC members. The particular focus of Africanist discontent was the Preamble of the Freedom Charter, which declared:

> That South Africa belongs to all who live in it, black and white, and that no government can justly claim authority unless it is based on the will of the people; That our people haven been robbed of their birthright to land, liberty and peace by a form of government founded on injustice and equality; That our country will never be prosperous or free until all our people live in brotherhood, enjoying equal rights and opportunities.

Following the Congress of the People and the publicization of the Charter, Hezekiel Ndaba, a resident of Moroka Camp outside Johannesburg, issued his own pamphlet entitled "An African Reviews Freedom Charter." Whether Ndaba considered himself an Africanist or not is unclear. He was never a prominent Congress member. Yet his pamphlet set out clearly the problems Africanists considered central in their dismissal of the Charter. Ndaba suggested that the acceptance of the Charter implied both a change of policy and heart for members of the liberation organizations, a contention of the Africanists who believed that the Charter was a clear departure from the ANC's Programme of Action.

Taking up the contentious line of the Preamble that spoke of South Africa as belonging to all, Ndaba declared that South Africa had "practically been made safe for whites alone, through conquest and trickery. It is exclusively preserved for them by force. As matters stand today, we have no share in its wealth, and play no part in its ownership. We have been deprived of all that belonged to us."[50] Ndaba's point of view differed, at least rhetorically, from that of the Africanists, for his particular indictment of the Charter was addressed not specifically to its substance but to the perceived gap between the Charter's vision and the realities obtaining and the Charter's failure to differentiate between the two:

> When the Charter says: "We, the people of South Africa, black and white together -- equals countrymen and brothers adopt this Charter" it merely expresses an objective desire on the part of its drafters. Yes, we stand for equality and brotherhood of man, but we cannot be bluffed. We are certainly not equals in South Africa, and we must say so. Let the good Charter express the aims of which the few 'Progressives' support, and not to put it as a reality because it is misleading.

> When the Charter says: "Our people have been robbed of their birth-right to the land," it is perfectly correct only as far as the blacks are concerned. Therefore not until there will be the same laws for both black and white South Africans, and equal opportunities for all, will there be any right for anybody to claim that there is equality between the races of this country.[51]

There was a hard edge to Ndaba's statement, a sense that the Charter had been drafted by and expressed the perceptions and will of "the few 'Progressives,' " those white "brothers" who could speak of equality because they were members of the minority that had set the terms of inferiority and enjoyed the benefits of supremacy. Ndaba spoke to the crux of the controversy incarnate in the Freedom Charter. The central problem was that of how the relationship between a future non-racialism and the immediate need for the attainment of African power were to be mediated through the organizations of the national liberation movement. How was one to square a desire for a society based on mutual understanding, cooperation and respect, and without reference to race, with the immediate needs and apparent realities of struggle whereby a mass of indigenous citizens dispossessed of their rights, their personal property, and their history, were attempting to organize among themselves to realize their birthright? Ndaba was concerned that Africans would lose sight of the will and right to power which had to be the fundamental force motivating the liberation struggle. He feared that Africans were expending too much of their energy in alliances, the product of which was a Charter that neglected the question of who was struggling and for whom in favor of a vision presented as if it were already a reality. Therefore, the Charter, far from ameliorating the tensions and latent divisions among the African political community, set the terms by which the battle for the hearts and minds of Africans would be contested.

The difference between Ndaba and the Africanists was the latter's willingness to spell out clearly, forcefully, and directly their critique as it related to the Congress Alliance, the presence of whites in the national liberation movement, and the then-current ANC leadership.[52] The Africanists considered themselves orthodox nationalists, the keepers of the true African nationalist flame in the face of the ANC's turn to unfruitful ideologies and foreign tutelage. The Africanists agreed with Ndaba that the African had been robbed of his land, but, unlike Ndaba, they were completely unwilling to abide any alliance between Africans and others that might confuse the aims of African nationalist struggle in the minds of the African people. The Africanists believed that the national struggle was primarily psychological. Victory was a question of raising consciousness and mobilizing, renewing the African's faith in her or his own history and selfhood, and presenting a completely unambiguous picture of the necessary steps to be taken to achieve freedom. Therefore, alliances with whites or Indians served, at best, only to confuse the masses, and, at worst, such alliances continued to propagate the apartheid tenet that Africans could do nothing for themselves. In the Africanist mind white radicals and white liberals were of a piece. Both were concerned only to protect their narrow sectional interests. The Africanists perceived the non-racial socializing of the 1950s as a means by which the African leadership was lured into a sense of comradeship and equality with foreigners, particularly whites, who were then able to win important positions in the Congress Alliance, or, in the case of Liberal patronage, draw the African away from more radical forms of struggle.[53] Whites, particularly those in the Congress of Democrats, were emasculating African nationalism and placing in its stead multi-racial visions without reference to the mobilization of the African majority:

> Even if we grant the sincerity of the whites, Indians and Coloureds who want to collaborate with us, the fact remains that the only way in which white domination will ever be broken is by black force. When that day comes if we have to stop and ask ourselves whether a particular white man was a friend of ours in the past, then we will never be able to act. After it is all over we will grant all those who accept African hegemony their full rights as private citizens of an African State.[54]

It is understandable, then, that to the Africanists, the Freedom Charter was evidence of the success of whites and other "foreign elements" in pushing their ideological program to the forefront within the Congress Alliance and the ANC. Potlako Leballo, perhaps the most vocal Africanist rhetorician, called the Charter a "political bluff" and demanded that the ANC return to the Programme of Action. Leballo found nothing prescriptive in the Charter and stated that it was useless "to go around shouting empty slogans such as 'The people shall govern,' 'The people shall share,' without practical steps towards that government."[55] The Africanists were saying the same thing that Ndaba had addressed, but in much less diplomatic language, and to different ends. Both feared the imposition of an agenda from without that would divert the ANC from its appointed task: the winning of freedom and its attendant material and psychological benefits for the African people. Both were concerned that the question of power had become secondary, in favor of questions of organization and tactical and material needs, and would be lost in the tangle of these and other concerns. Both recognized also that the leadership of the African people could be lost to them if the people became confused, or, as Ndaba put it, if the leaders continued to "[build] castles in the air" rather than to take action. Their differences arose in Ndaba's apparent unwillingness to dismiss collective action, represented by the Congress Alliance, outright. The Africanists did so consciously and in the belief that to do otherwise was to endanger the struggle itself.

Leaders of the ANC spoke up in defense of themselves, the Congress of Democrats, the Congress Alliance, and its Freedom Charter. Both Dan Tloome and Walter Sisulu wrote lengthy articles defending Congress, and both described the Africanists' doctrine as "inverted racialism." Tloome took the Africanists to task for propagating "lies," particularly those stating that COD controlled and dominated the ANC. He congratulated the African working class for their maturity in understanding "that if people of Indian, Coloured or

European origin are prepared to come forward, honestly and without reservations, as allies in the struggle for these things [freedom, equality, a living wage], such friendship must be welcomed with both hands."[56] In the second part of his series on Congress and the Africanists, Tloome disputed that the Africanists had a meaningful ideology. Stating that in some circumstances, " 'Africa for the Africans' " [was a] "sound, militant and correct slogan," Tloome suggested that the slogan, in its correct usage, referred to foreign invaders and colonialists but not to those who lived in South Africa and knew no other home. Tloome accused the Africanists of using the slogan in a singularly racial manner, so obscuring its meaning. Tloome acknowledged the wicked and unjust social and political system in South Africa and posed the Freedom Charter as a solution or blueprint for change, while suggesting that Africanism would simply continue the same pattern of injustice. Africanism would "weaken the overwhelming moral justice of our cause." In his concluding remarks, Tloome called the Africanists "so-called Nationalists" because they sought to perpetuate and accentuate differences just as Verwoerd and Swart did.

Why did the leaders of Congress, including seasoned political leaders like Sisulu and Tloome, reserve some of the harshest words they would ever express, words as harsh as any meted out to the Nationalists, for the Africanists? There are a number of possible explanations. It could be that the ANC saw the Africanists as posing the single greatest threat to it. The anger of ANC leaders may have been in direct proportion to those leaders' recognition of the prospective drawing power of Africanism. Even Sisulu was forced to admit that the climate in South Africa promoted increasing hostility to whites. Sisulu acknowledged that "a policy of extreme nationalism" might enjoy a degree of acceptance.[57] Perhaps the degree of anger expressed was a result of ANC leaders' sense of responsibility toward their allies in the other Congresses. The Africanists' rhetorical barrage, no matter how crude it may have appeared to Congress leaders, was a direct and frontal assault on the ideological structure that the Congress Alliance had built and capped with the Freedom Charter. The ANC was the leader of that Alliance and the primary legitimating element therein. Therefore it fell to the ANC to take the offensive against the Africanists, and that they did with a vengeance. For this reason, it is doubtful that the statement below, mild and patient in tone, published as part of an editorial in *Liberation*, was written by an ANC member:

> What is this "Africanism" everybody is talking about? It would not be unfair to describe it as a sort of "nationalism" of the Malanite variety, turned upside down....Now it is true that Congress has got beyond this immature viewpoint to the deeper wisdom of rejecting each and every variety of master-race ideology in favor of a truly human vision of equality and brotherhood. But isn't it also true that the very fact of oppression and the official preaching of White

> *baaskaap* inevitably brings forth this very reaction of African assertiveness and exclusiveness? In a sense, this is the necessary first step towards casting out the servile mentality inculcated by the ideologies of 'Bantu' inferiority. It has always existed in Congress and will continue to exist, under whatever name, as the Congress continues to recruit new members, welcome them and educate them; not drive them away by swearing at them! The "Africanists" of today -- all but a handful of cranks who will soon expose themselves -- are the good and loyal Congressmen of tomorrow: we have seen all this happen before.[58]

If this attitude was representative of a more generalized outlook in the Congress Alliance, it was rarely in evidence. The ANC was taken to task for "swearing" at the Africanists and their supporters. The emphasis on education through experience in the ANC and the sense of inevitability that Africanists would ultimately become loyal Congress members was a marked departure from any publicly expressed viewpoint of an ANC leader. Africanism was considered to be a normal stage of growth in a unilateral progression toward political maturity. The author was undoubtedly thinking of the ANC leaders, Nelson Mandela among them, who were weaned on the African nationalism of the 1940s and subsequently dropped their parochial outlooks in favor of multi-racial cooperation and a non-racial vision:

> For the first time in the history of our country the democratic forces irrespective of race, ideological conviction, party affiliation or religious belief have renounced and discarded racialism in all its ramifications, clearly defined their aims and objects and united in a common programme of action.[59]

Africanism was not simply an ideology of "cranks," nor was it to be so easily dismissed. The mistake made in dismissing it was to confuse the messenger and the message. Even though African leaders such as Potlako Leballo and Josias Madzunya were dismissed as cranks, the power of core Africanist ideas had not proven to be bankrupt. And how were the observers of Congress to interpret the adoption of the Freedom Charter, or the Charter itself? The Congress of the People had been a celebration of a process that involved ordinary South Africans in speaking for themselves and coming together in an alternative parliament. This was a unique and moving achievement. But had the product, the substance of that process, brought about a common understanding among the participants? Did the Freedom Charter represent the end of a struggle to define terms such as "national," "liberation," and "non-racialism"? Or rather was the Charter the first concrete expression, by the Congress Alliance members, of these questions? Clearly the Congress and the Charter emerging from it raised a number of highly contentious issues within the national liberation movement. The Africanists challenged the basic legitimacy, not of the ANC proper, but of the ANC leadership and the alliances forged by it. In promoting the Congress Alliance and the Freedom Charter, the ANC leadership had chosen consciously to move away from the orthodox nationalism of the Africanists. This made the ANC vulnerable because ANC leaders could understand the power of Africanist doctrines but did not share similar tactical or ideological visions. The leadership thus found itself in the position of attacking the Africanists as racists while defending ANC alliances with non-Africans.

After Congress: The Million Signatures Campaign

Following the Congress of the People, the Charter was referred to the Congress affiliates for discussion, emendation, and ratification. The process was a lengthy one within the ANC, and there were some who felt that there had been too little substantive discussion before the Charter was ratified. A new consultative body emerged from the Congress of the People campaign. The National Consultative Committee (NCC), composed of members from each of the four Congresses and SACTU, was to oversee general cooperation between the Congresses and, specifically, the Million Signatures campaign, a petition drive to familiarize the public with the Freedom Charter. Echoing the hopes for the Congress of the People, the National Action Council wrote in its final report that it expected the Million Signatures campaign to move the Congress affiliates to "go out amongst the people and establish contact with them on an unprecedented scale."[60] The campaign was unsuccessful in reaching its goal. The target date for collection of a million signatures gave the Congresses less than a year to complete the task. Reporting back to the Congress affiliates shortly after the deadline date for the campaign, the NCC noted that less than ten percent of the desired signatures had been collected. The NCC report complained of a number of problems, including poor administration of the campaign, poor communication between the NCC and its provincial and regional counterparts, and underfunding.[61]

The Congress of Democrats responded vigorously to the Million Signatures campaign, seeing it as a means of introducing COD and the Congress Alliance to the white public while collecting signatures and recruiting new members.[62] It was envisaged that COD members would distribute leaflets and circulate petitions in as many venues as possible, including door-to-door work, collection at factories and trade unions, and various types of clubs and organizations. Freedom Charter work would be combined with specific issues if at all possible.

Most of COD's signature collection took place on the streets of cities, where members set up tables. *CounterAttack*, COD's internal publication, reported on the Hillbrow branch's progress in signature collection. The branch was to call a meeting of Charter signatories to acquaint them generally with

COD's policy. The article congratulated the branch for its "correct understanding" of the signature campaign and its potential for expanding COD and bringing the Congress Alliance to whites. Written approximately six months after the initiation of the campaign, the article noted that nearly 1,000 whites had signed the Charter.[63] An article three months later continued praise for Hillbrow's efforts, to the effect that: "Several hundred people signed their endorsement of its principles at our tables. We always explained carefully that the Charter demands equal rights for all races, and many signatorees [sic] make remarks such as, 'About time too!', 'They are also human beings!' and so on."[64] There were follow-up visits to signatories in their homes, and COD members were generally well-received, finding people to be willing listeners and discussants. However, this activity did not bring new members to COD. Few of the signatories came to the group meetings to which they had been invited. Michael Picardie was involved in the collection of signatures for the Freedom Charter and remembers table-sitting in the white suburbs of Johannesburg:

> It must have been some time in autumn, April, May. I would turn up on a Saturday morning in Bates Street, Doornfontein, which is between Hillbrow and the city, and we set up a table outside a delicatessen shop called Crystal's, a Jewish delicatessen, and we had a table, and we had a petition, asking people to sign the Freedom Charter....And people were just astonished and amazed, and bewildered by it. I mean, what were these Jewish housewives to make of [this], and just ordinary Afrikaner whites, working-class whites living in Doornfontein, what were they to make of the Freedom Charter? Anyway, the inevitable happened, and I was arrested one Saturday morning at the table.[65]

This was the rule rather than the exception for those sitting at the Charter tables. It is illustrative of South African political and cultural mores at the time that while an African caught without a pass could take for granted a fine or more probably a jail sentence, Michael Picardie was driven around by the police, scared a little and then brought home without further penalty for distributing a document considered subversive and communistic. In the mid-1950s whites, even leftist ones, could expect a degree of consideration from the police and Special Branch never extended to a black person. COD's Cape Town experience was similar to that of Johannesburg. Tables were set up in Sea Point and Claremont, and the latter site's table was raided by the Special Branch after only 20 minutes, and its documents confiscated. The Sea Point table enjoyed greater longevity and some success, collecting "a good number of signatures" and distributing leaflets.[66]

Writing their report for the annual national COD meeting, the Cape Western Regional COD Committee noted the establishment of "healthy relations" with the ANC and SACPO. Members of COD were serving on the Cape Western Consultative Committee, and the author of the report noted that

COD had been "instrumental" in the furtherance of the Million Signatures campaign. The report also contained criticisms of failures due to poor organization and lack of enthusiasm. The National Consultative Committee was criticized for its insufficient propaganda work around the Charter. In particular the Western Cape COD stated that it was unable to get English language copies of the Charter.[67] There were many organizational problems within and between the Congress affiliates. Still, it is unlikely that COD's failure to collect whites' signatures was a reflection of poor organization or the lack of English language Charters available. As Picardie described it, whites, even in the neighborhoods considered most progressive in Johannesburg and Cape Town, were generally unable to comprehend the Charter. COD had proved its willingness to work hard on every aspect of the Congress and Charter campaigns, but there were limits. A report on the Durban branch's work noted the branch's discouragement at their first Charter work among whites. Also, the group was involved with the other Congresses at the time in anti-pass and Group Areas campaigns. There was only so much energy that could be expended on difficult white area work, and the realities of Nationalist initiative dictated that energies would shift to the more immediate concerns of the people. In the end, the NCC admitted the failure of the Million Signatures campaign, and the primary reason, in the Committee's estimation, was the inability to link the Charter to everyday problems, coupled with a tendency to "treat the Charter as a noble dream."[68]

The Treason Trial: Solidifying Congress Identity

The police raids of early December 1956, netting 156 Congress Alliance leaders and members, was the beginning of a judicial nightmare that would last through the end of the decade and into 1961. The charge was high treason, the state suggesting that those participating in the Congress of the People and disseminating the Freedom Charter planned to overthrow violently the South African government. The arrestees were brought to Johannesburg and kept in the infamous Fort for 16 days. The Preparatory Examination lasted over a year, after which 64 defendants were excused. The remaining 91 were charged with high treason. Finally, the indictment was narrowed to include a core of 30 defendants, representing all of the Congresses, and these defendants were acquitted in the fifth year of the proceedings, on March 29, 1961. The events of the trial procedure itself are well-documented.[69]

The effect of the Trial on those participating in it, as members of sister organizations and as individuals, was probably a more important aspect of the Trial than the acquittal judgment. Chief Lutuli, writing in the introduction to Helen Joseph's book on the latter half of the Treason Trial, stated that "In all humility I can say that if there is one thing which helped push our movement along non-racial lines, away from narrow, separative racialism, it is the Treason Trial, which showed the depth of the sincerity and devotion to a nobler cause on the white side of the colour line."[70] Lutuli's statement is interesting, as it suggests that prior to the Treason Trial, when the Congress of the People and the Million Signatures campaign concluded, the Congresses' mutual distrusts and misapprehensions remained. Lutuli suggests that the Treason Trial proved to be an acid test in the matter of cooperation and deep mutual trust. Rusty Bernstein, echoing Lutuli, considered the Trial to have been a key factor in realizing a true non-racial spirit in the Congress Alliance and creating the strong personal relationships between black and white that would bridge the mass activism of the 1950s and the 1980s:

> Something that interests me was the fantastic implication of the Treason Trial in 1956. Not just for its political value, but that the government brought together 156 people who scarcely knew each other because of the sort of provincial separation in South Africa. I mean, we might have known each other vaguely, or met at a conference five years ago. They [the government] brought [us] together and they stuck us in a courtroom every day six hours a day for years on end.[71]

Bernstein considered it quite ironic that the state was responsible for putting the far-flung members of the Congress Alliance in a position to integrate the liberation movement nationally:

> There was no segregation, because it was done alphabetically....The only place in South Africa where there was absolutely no color bar was in the Treason Trial. And there we were all stuck together, so apart from making this tremendous contribution to uniting the South African movement from what had really been a federated body, and with three or four different organizations, a racial body, it produced a core at the center of people who really thought of themselves as a united group. I mean, the Treason Trial [group] was a sort of family of their own at the end of three years.

Bernstein was one of a number of individuals who noted that there was a surge of non-racial socializing around the Treason Trial which was an important aspect of melding the core to which Bernstein refers:

> [The Treason Trial] also established this sort of easy-going social relationship, I think, between blacks and whites which hadn't existed previously, even among the politicals. We did socialize [before] but we were rather terribly nervous of each other because we were all conscious of the fact that we were doing something odd. But here, after three years of eating lunch together, singing songs together, sitting together, exchanging crossword puzzles and so on, you really had a feeling for the first time that here were people, it was just another social group, you weren't conscious of what color or race or anything

> people were. And I think really that the Treason Trial, in a way, when looked back on, was the most important single factor in breeding the present South African movement. I think the government there created a Frankenstein without knowing it. We didn't even realize at the time what it was doing, but it transformed the whole character of the South African movement from that time on.

The Treason Trial may have provided the first environment, as Bernstein suggests, in which blacks and whites who had been friends and co-workers before were together for a sufficiently long time, sharing the same frustrations and hardships, to truly break down the barriers of race, class, and language and the complexes of inferiority and superiority that hindered the realization of a truly unified core group at the heart of the Congress Alliance. However the value of the Treason Trial to the national liberation movement had an organizational component as well:

> The leadership certainly came to know each other quite well for the first time. We got to know the top cadres throughout the country quite well during that period and that was important. It welded together the Alliance, hearing each other's speeches in court and all that kind of thing gave us a perspective on each other's movements.[72]

Denis Goldberg remembered that throughout the Treason Trial Moses Kotane, a member of the ANC but more centrally a leading member of the Communist Party, sat next to Chief Lutuli.[73] Of course, the Treason Trial was a time of great hardship for many of the defendants. With the defendants sequestered in Johannesburg, families were separated, jobs were lost. The very boredom of sitting in court year after year was a trying experience. Yet, in terms of creating a united, non-racial opposition, the Treason Trial was a very useful investment in the future of the national liberation movement. Liberals' unwillingness to make such an investment guaranteed their marginalization. Robertson, as noted earlier, has suggested that the work of Liberal Party members in the Treason Trial Defence Fund and other support activities brought the ANC and the Liberal Party closer together than they had been since the Liberals bowed out of the Congress of the People. Robertson points specifically to the opportunities for social contact between liberal whites and the leaders of the Congresses, as those on trial were staying in the Johannesburg area.[74] Perhaps, in the intervening years, some Liberals had rethought the validity of extra-parliamentary action and had therefore moved closer philosophically to the Congress point of view. Although the Liberal Party won praise for its aid in the Treason Trial period, and did enjoy a closer relationship with the ANC to the end of the decade, the party stood on the outside looking in at the fortified Congress core. If Liberal Party members had been among those being tried for treason, it might have been possible to create a working Congress-Liberal Party relationship. But the party had chosen to stand on the outside of this crucial formative experience, and though, as discussed in the next chapter, the Liberal Party and the Congress affiliates would meet in a number of arenas in the 1950s to hold discussions and at times even to work together, the Liberal Party was destined to remain on the outside.

The Congress of the People divides the 1950s chronologically but suggests other divisions as well. From that moment forward the African National Congress and its affiliates could point to a Charter, a document that was a lodestar of sorts, though hardly a blueprint, for a post-apartheid future. Still, it was hard copy, and one could point to aspects of it and say, "this is my vision also" or "no, I do not believe this and I do not accept this." As such, it could be taken as an inspiration to Congress members who accepted it or as a challenge, even an instigation, to members, like the Africanists, who did not. The Charter's first substantive declaration trumpeted that "South Africa belongs to all who live in it, black and white." For all those involved in the national liberation movement, and particularly for the members of COD, this was a critical and momentous declaration. The ANC had publicly declared that all South Africans who accepted the Charter's precepts and vision were equal as South Africans and would be so treated in a new South Africa. Thus, the national liberation movement had been declared, by its African leadership, to be a struggle not against whites as oppressors, but rather against an evil system -- a struggle in which all right-thinking South Africans could involve themselves. And as a corollary, the new South Africa would not be defined by racial and ethnic majorities and minorities but by a commonly defined standard of citizenship, which would guarantee that no one would have to feel a second- or third- or fourth-class citizen. This vision was unacceptable to Africans who believed wholeheartedly that, both in struggle and in a post-apartheid South Africa, African status would be a function of African paramountcy. Defining the national liberation movement and post-apartheid society in a colorblind manner was thus ceding African primacy, power, and moral authority, and seriously endangering the construction of an African identity that would infuse Africans with the will to resist. From the Congress of the People and its child, the Freedom Charter, the battle lines were drawn that would remain manifest in the national liberation movement up to the present.

Notes

1. Raymond Suttner and Jeremy Cronin, *30 Years of the Freedom Charter* (Johannesburg: Ravan, 1985), 196.

2. Thomas Karis and Gail M. Gerhart, *From Protest to Challenge: A Documentary History of African Politics in South Africa, 1882-1964*, Vol. 3, *Challenge and Violence, 1953-1964* (Stanford: Hoover Institute Press, 1977), 57.

3. South African Indian Youth Congress, "Listen Young Friend," pamphlet, n.d., in Suttner and Cronin, *30 Years*, 21.

4. Suttner and Cronin, *30 Years*, Interview with A.S. Chetty, 47-48.

5. Karis, Carter and Gerhart, Vol. 3, *Challenge and Violence*, 61.

6. Douglas Irvine, "The Liberal Party, 1953-1968" in Jeffrey Butler, Richard Elphick, and David Welsh, eds., *Democratic Liberalism in South Africa: Its History and Prospect* (Middletown, CT: Wesleyan, 1987), 127, and Janet Robertson, *Liberalism in South Africa, 1948-1963* (Oxford: Clarendon Press, 1971).

7. Robertson, *Liberalism*, 165-167.

8. Ibid., 161.

9. Dave Everatt, " 'Frankly Frightened' ": The Liberal Party and the Congress of the People," unpublished paper presented to the "South Africa in the 1950's" Conference, Queen Elizabeth House, Oxford, September 1987.

10. Ibid., 8.

11. Ibid., 18.

12. Ibid., 20.

13. Arden Winch, Liberal Party memo on the Congress of the People Banquetting Hall Meeting, 18 August 1954, quoted in Everatt, 22.

14. Everatt, " 'Frankly Frightened,' " 24.

15. Ibid., 30.

16. Karis, Carter and Gerhart, Vol. 3, *Challenge and Violence*, 58.

17. Congress of Democrats, "For Discussion and Consideration by Congress of Democrats Regions and Branches, and a Guide for Future Activity," n.d., Treason Trial Documents, CAMP Microfilm 405, Reel 19.

18. *Fighting Talk*, "The Congress of the People: Will the Europeans Be There?" April 1954.

19. Congress of Democrats, "Addendum to Minutes of Transvaal Congress of the People Committee, 22 February 1955, all present, 'Report of COD Representative to Provincial Committee of Congress of the People, 22nd February 1955,' " Treason Trial Documents, CAMP Microfilm 405, Reel 7.

20. Suttner and Cronin, *30 Years*, 71.

21. African National Congress, T.E. Tshunungwa, National Organiser, ANC and Congress of the People, "Report of My Visit to the Cape Western Region ANC branches, January 1955," Treason Trial Documents, CAMP Microfilm 405, Reel 1.

22. Correspondence, Cape Western Action Council, Congress of the People, to National Action Council, 5 April 1955, Treason Trial Documents, CAMP Microfilm 405, Reel 7.

23. The committee membership as a whole was composed of five members from each of the sponsoring organizations and another fifteen elected at a conference which had been held at the time of the campaign's initiation.

24. Correspondence, M.B. Yengwa to Dr. W.Z. Conco, 17 June 1955, Treason Trial Documents, CAMP Microfilm 405, Reel 11.

25. Lodge, *Black Politics*, 71.

26. Transvaal Provincial Committee of the Congress of the People, "Report of Treasurer to the Provincial Committee of the Congress of the People, 22 February 1955," Treason Trial Documents, CAMP Microfilm 405, Reel 7. COD's Hillbrow branch collected over £27 of the total.

27. Transvaal Provincial Committee of the Congress of the People, Minutes, 24 May 1955, Treason Trial Documents, CAMP Microfilm 405, Reel 7.

28. Minutes, Transvaal Provincial Committee of the Congress of the People, 25 January, 22 March, 29 March, 5 April, 12 April, 26 April, 3 May, 10 May, 17 May, 24 May, 1955, Treason Trial Documents, CAMP Microfilm 405, Reel 7.

29. Correspondence, T.E. Tshunungwa to Secretary, Congress of Democrats, Cape Town, 12 September 1954, Treason Trial Documents, CAMP Microfilm 405, Reel 20.

30. Meeting, ANC, SACPO, TIC, SACOD, Congress of the People Conference, 25 July 1954, Treason Trial Documents, CAMP Microfilm 405, Reel 9.

31. Congress of Democrats, "Notes on the Political Situation by the National Executive Council for Discussion at Conference," n.d. Treason Trial Documents, CAMP Microfilm 405, Reel 14. Though undated, this report was written for the COD annual conference which was held immediately prior to the Congress of the People.

32. Document, n.d., n.a., Treason Trial Documents, CAMP Microfilm 405, Reel 23.

33. Karis and Gerhart, Vol. 3, *Challenge and Violence*, "Police record of the Congress of the People, Kliptown, Johannesburg, June 25-26, 1955," 202.

34. Ibid., 203.

35. Nelson Mandela, "In Our Lifetime," *Liberation*, No. 19, June 1956.

36. Interview, Rusty Bernstein, Dorstone, 25 January 1987.

37. Draft Freedom Charter, n.a., n.d., Congress of the People File, Popular History Trust, Harare.

38. Suttner and Cronin, *30 Years*, 262-266.

39. Draft Freedom Charter.

40. Copies of "The Women's Charter" and "The women's demands for the Freedom Charter" can be found in Suttner and Cronin, *30 Years*, 161-170. I do not know if there were women on the drafting committee.

41. Suttner and Cronin, *30 Years*, 263.

42. Interview, Ben Turok, London, 27 January 1987. Turok introduced the economic clause on the second day of the Congress, 26 June 1955. For the police transcript of his introduction, see Karis and Gerhart, Vol. 3, *Challenge and Violence*, 194-195.

43. Ibid.

44. Ibid., 64.

45. Ibid., 71.

46. Michael Harmel, "After the Colonial Revolution," *Liberation*, No. 25, June 1957, 15-16.

47. Nelson Mandela, *The Struggle is My Life* (London: International Defence and Aid Fund, 1986), 173.

48. H.J. Simons, "South Africa's Power Structure," *Fighting Talk*, Vol. 15, No. 2, March 1961.

49. Form letter to Africans in rural areas, beginning, "You must have heard that there is going to be a very big 'indaba' in Johannesburg on Saturday 25th June and Sunday 26th June, 1955," n.a., n.d., Congress of the People File, Popular History Trust, Harare.

50. Hezekiel Ndaba, "An African Reviews Freedom Charter," n.d., Congress of the People File, Popular History Trust, Harare.

51. Ibid.

52. This discussion owes much to Gail Gerhart's work, *Black Power in South Africa: The Evolution of an Ideology* (Berkeley: University of California Press, 1978).

53. Gerhart, *Black Power*, 157.
54. *Contact*, publication of the South African Liberal Party, 8 March 1958.
55. P.K. Leballo, "The Nature of the Struggle Today," *The Africanist*, December 1957.
56. Dan Tloome, "The Africanists and Congress," *Fighting Talk*, August 1958.
57. Walter Sisulu, "Congress and the Africanists," *Africa South*, July-September 1959.
58. Editorial, "Searchlight on the Congresses," *Liberation*, No. 29, February 1958.
59. Nelson Mandela, "Freedom In Our Lifetime," *Liberation*, June 1956.
60. South African Indian Congress Agenda Book, 1956, Annexure B.3, "Report of the National Action Council of the Congress of the People to the Joint Executives of the A.N.C., S.A.I.C., S.A.C.P.O. and S.A.C.O.D.," Ginwala Papers, Institute of Commonwealth Studies.
61. "Report of the National Consultative Committee Presented to the Joint Executives of the A.N.C., S.A.I.C., S.A.C.P.O., S.A.C.O.D. and S.A.C.T.U.," Ginwala Papers, Institute of Commonwealth Studies.
62. Congress of Democrats, "National Executive Committee Plan for the Campaign to Popularize the Freedom Charter and the Congress of Democrats," n.d., Treason Trial Documents, CAMP Microfilm 405, Reel 22.
63. *CounterAttack*, March 1956, Treason Trial Documents, CAMP Microfilm 405, Reel 2.
64. *CounterAttack*, June 1956, Treason Trial Documents, CAMP Microfilm 405, Reel 2.
65. Interview, Michael Picardie, Cardiff, 24 January 1987.
66. *CounterAttack*, June 1956.
67. Congress of Democrats, "Cape Western Region, Draft Report," n.d., Treason Trial Documents, CAMP Microfilm 405, Reel 12.
68. *CounterAttack*, excerpt from the National Consultative Committee, "A Report of the Freedom Charter Campaign," July 1956. Some criticisms were more pointed and more serious in their implications. In the February 1956 issue of *Liberation* the editorial scolded the Congress movement for letting down after the Congress of the People and not taking organization around the Freedom Charter with sufficient seriousness. The ANC was highlighted in particular for its 1955 Annual Conference, where the writer claimed hours were spent debating the question of whether to allow admission of the *Bantu World* reporter, while discussion was postponed on mobilization around the Charter until April.
69. See Anthony Sampson, *The Treason Cage: The Opposition on Trial in South Africa* (London: Heinemann, 1958), Lionel Forman and E.S. "Solly" Sachs, *The South African Treason Trial* (New York: Monthly Review Press, 1958), and Helen Joseph, *If This Be Treason* (London: Andre Deutsch, 1963).
70. Joseph, *If This Be Treason*, 9.
71. Interview, Rusty Bernstein, Dorstone, 24 January 1987.
72. Interview, Ben Turok, London, 27 January 1987.
73. Interview, Denis Goldberg, London, 3 February 1987.
74. Robertson, *Liberalism*, 181-182.

9

Building the United Front in the Late 1950s

The period between the Treason Trial arrests and the Sharpeville massacre in March of 1960 witnessed a change in the timbre of the national liberation movement. The ANC, lacking in direction after the Congress of the People and stunned by the massive arrests that took many Congress leaders out of commission, sought to build a broader front, encompassing liberals and religious leaders, against the government's onslaught. A series of multi-racial conferences were held in 1957 and 1958 bringing these diverse constituencies together in discussion. These conferences foreshadowed the alliances that gained momentum through the last years of the decade. The 1958 national election was a primary generator of that momentum. The National Party's victory, its largest ever, suggested that the party would not be unseated by parliamentary means in the near future. The Liberal Party was led to question the basic tenets of its faith, including parliamentarism, and in 1958 it advocated boycott as a vehicle of protest. By mid-1960 the party was advocating universal suffrage.[1]

The Congress of Democrats (COD) was also undergoing changes in these years. The majority of COD's leadership had been banned by, or during, the Treason Trial. A new group of leaders rose to take those positions, a group neither particularly experienced in leadership, nor as ideologically radical, as many of its precursors had been. Still, there was a diversity of strong leftist voices within COD through the end of the decade, and the clash of these and more moderate voices precipitated an internal crisis in 1958. COD's membership was divided between those calling for renewed efforts to engage the working class and those demanding that COD de-ideologize itself to attract liberals. The more moderate voices seeking accommodation with liberals carried the day.

The relationship between COD and the ANC remained generally strong through the end of the 1950s. COD, following the ANC's lead, worked with a variety of liberal organizations and promoted the ANC's leadership and program among liberal whites. However, a tension, present throughout the life of this relationship, was apparent in the last years of legal Congress Alliance activity. This tension was the product of differing understandings between COD members and their ANC counterparts as to the primary focus and purpose of the national liberation movement. In the late 1950s a number of whites, some in COD and others without organizational affiliation, promoted the fusion of the Congress Alliance affiliates in a single organization. This controversy was known as the "One Congress" question.

The raising of the question was a matter of embarrassment to the leaderships of both the ANC and COD. The ANC was forced to remind the COD membership that the struggle was first and foremost for the purpose of winning African liberation: The creation of a non-racial society was desirable but not primary. Whites' raising of the question suggested anew the ambiguity of the white presence in a struggle for majority liberation from white domination. The maintenance of the African National Congress (ANC) as a separate African organization was elemental in the minds of Africans. That sympathetic whites, particularly those in COD, could not see this may have led the ANC to maintain a certain skepticism regarding COD. COD had proved itself a trustworthy ally by the late 1950s. However, it was still not necessarily clear that democratic whites and politicized Africans shared the same terms of analysis. In the last years of the 1950s COD members sent memoranda to Pan-African conferences and applied for affiliation to Pan-African organizations. COD hoped to offer the uniqueness of multi-racial cooperation in the South African struggle as an example to other African states in the process of de colonization. Yet in South Africa multi-racialism and non-racialism were still very much experimental. There were no precedents for the Congress Alliance, and it had yet to be seen if such alliances were sustainable.

Broadening the Front: The Multi-Racial Conference

In a letter to Prime Minister Strijdom in 1957 Chief Lutuli bemoaned the lack of consultation between blacks and whites and criticized the existing structures for such contact as inadequate. Lutuli pointed to the lack of such structures as central to the "growing deterioration in race relations" and called for a multi-racial convention as a means to address the national crisis.[2] Lutuli received no satisfaction from the government. However, in late 1956 the Interdenominational African Ministers Federation (IDAMF) sponsored a multi-racial conference, and as a result of this effort the ANC called for a further conference with the purpose of building a more diverse multi-racial united front. In 1957 and 1958 a number of multi-racial conferences were held under the auspices of African religious leaders and various black and white, radical and liberal, sponsors. The most critical and well-attended of these conferences was the Multi-Racial Conference at the end of 1957. ANC members participating included Professor Matthews, Chief Lutuli, Oliver Tambo, and Duma Nokwe. Was the ANC's promotion of these conferences and the building of a broad front a signal to the other Congress affiliates to consciously broaden their contacts outside Congress? The ANC had sought to forge these links before, but from the time of the IDAMF conference there was a clear, renewed focus on the united front.

The Multi-Racial Conference held in the Great Hall of Johannesburg's Witwatersrand University at the end of December 1957 resulted from IDAMF's inviting a number of prominent whites to join in sponsorship. Among the Conference sponsors were Chief Lutuli, Bishop Reeves of Johannesburg, Yusuf Dadoo of the South African Indian Congress (SAIC) and South African Communist Party (SACP), Leo Marquard, a prominent liberal, and other religious and liberal community leaders. One thousand people were invited on an individual, rather than an organizational, basis.

During the conference's three days a number of prepared papers treating various problems facing a multi-racial society were read: topics included the responsibilities of religious communities, educational policy, economic and civil rights and political arrangements. Following these presentations, the assembled broke into smaller "commissions" charged with preparing "findings" in each of these areas.[3] The commissions' "findings" were general but strongly condemnatory of apartheid and its effects. The conference's greatest impact upon the participants was a function less of its production than of the process itself:

> Educationists, sociologists, professors, churchmen, trade unionists, politicians and others shared in the discussions in a wider circle than they had previously known. Anglicans, Roman Catholics, Methodists and other Christians met in discussions on the various topics with Jews, Moslems and Hindus. Members of the four Congresses were thrown together with Liberal, Labor and other party stalwarts.
>
> Often the discussions in the Commissions were lively. Several of the participants declared afterwards that the debates in the Commissions had been a new and a profound experience.[4]

Conference leaders noted that many of those in attendance believed their groups "stood at the crossroads," faced with the choice of cooperation or irreconcilability and the aftermath of that irreconcilability. The Multi-Racial Conference had opened up new lines of communication and had set a basis for discussion based on the findings of its commissions. The findings had called for universal adult suffrage on a common roll and the calling of a National Convention to frame a new constitution.

The ANC had not initiated these conferences, yet it embraced them and issued a call for the transformation of conference-bound unity into a political force. What moved the ANC to do this? Could this embrace of liberal and religious elements be interpreted as a retreat from African nationalist principle? Did it represent an abandonment of extra-parliamentary struggle in favor of a broad political alliance that might be able to influence whites sufficiently to turn the Nationalists out? The treason arrests had dealt a blow to the ANC and its partners, and the ANC engaged in little mass activism for the three years after the Trial. The most probable explanation for the ANC's outward-looking policy would understand it as a period in which ANC weakness, and a combination of liberal distress and increasing liberal politicization, conjoined. At a time when government repression was increasing, the possibility that the lack of communication between blacks and sympathetic whites could feed hatred and rebellion drew both camps together. There was much they could agree on by the later 1950s.

More surprising perhaps was the ANC's willingness to engage in dialogue with whites well to the right of the liberal camp. While the IDAMF and Multi-Racial Conferences were taking place, the South African Bureau of Racial Affairs (SABRA), an organization composed primarily of Afrikaner intellectuals, was considering holding its own multi-racial conference. SABRA's chartered mission was to prepare the country for fully realized "separate development." SABRA's *verligte* members sought to reconcile apartheid racial and ethnic divisions with Christian conceptions of social justice and equity, no small task. A SABRA committee was formed in late 1957 to consider the holding of a multi-racial conference. The conference never did take place. However, between December 1958 and February 1959 small groups of SABRA members engaged in private meetings with nearly 200 Africans. Among those involved were Chief Lutuli, Oliver Tambo, Nelson Mandela, Duma Nokwe, and the Africanists Robert Sobukwe, Potlako Leballo, and Zephania Mothopeng.

Jack Simons, a leading member of the South African Communist Party and a lecturer at the University of Cape Town, was not impressed by SABRA's words or actions: "What the academic intellectuals did, in addition, was to admit that continued White domination was immoral and transitory. That is, they 'discovered' something that radical and liberal thought had for long expounded, but that Afrikaners, as a body, had always vehemently denied."[5] Simons did not put a good light on the ANC's push to meet and discuss with organizations and individuals outside the liberation movement. He described the democratic camp as being "in the doldrums, and inclined to clutch at any sign of goodwill and friendship among government supporters."[6] The rounds of multi-racial communication continued through the end of the decade. In early

1959 Bishop Reeves of Johannesburg formed a consultative committee composed of representatives from fourteen organizations, including the Congresses, the Black Sash, and the Liberal Party. Meetings were held without agendas or written materials, giving the participants an opportunity to explore issues and come to an understanding. Discussion was thus sparked among organizations that had distrusted or possibly even feared one another. These meetings bore little fruit on the immediate political plane but were believed to be powerful in their ability to open eyes and break down barriers among culturally, politically, and intellectually disparate South Africans. Bishop Reeves, who closed the 1957 Multi-Racial Conference, was moved to remark that it might well be remembered in later years as "the turning of the tide in South Africa....Here we have demonstrated that it is still possible for those of various ethnic groups and holding widely divergent views to speak reasonably with one another. This is an achievement of which we may be justly proud."[7] Reeves' remarks suggest the power of those moments when the divide of color was bridged, a power of which the Congress of Democrats had long been well aware. Still, equally evident is the unwarranted optimism that could perceive such a moment as critical in defeating the Nationalists. If these moments were personally transformative, they were of little immediate value in the face of encroaching apartheid legislation and state force.

COD and Internal Crisis

A number of the resolutions put forth by branches at COD's 1958 national conference suggested that tensions in the organization were coming to a head, particularly in the wake of the Congress Alliance's attempts to build a broad front and embrace opportunities for joint activity with liberal and religious organizations. The Hillbrow branch's resolution was, at base, about keeping the organization's focus on primary oppression, that of class. The resolution stated that as the majority of Africans were workers, their "obvious place" would be in the trade unions and "their struggle would be along class lines" rather than along lines of race or ethnicity. The resolution, calling upon all COD members to work with the South African Congress of Trade Unions was a challenge to the COD leadership, and a declaration that united front activity would move COD and the Alliance even farther from the core ground of struggle, worker organization. The Hillbrow authors were repudiating the national-democratic liberation movement.

In stark contrast, the Congress of Democrats' Bellevue branch's resolution embraced the movement to forge links with organizations and individuals outside the Alliance, stating that since COD's 1957 conference the establishment of a "democratic, multi-racial society" had become increasingly acceptable to whites:

> This is evidenced by: a) The Multi-Racial Conference b) The stand taken by the Black Sash on the issuing of passes to women c) The progress in the Liberal Party's attitude on the franchise, and their greater willingness to co-operate with us d) The change in attitude of the press, which can no longer afford to ignore the Congress movement e) The large number of voters who supported our candidate in the Johannesburg municipal election.
>
> There is emerging among the Europeans a determined opposition to the Nationalists. The fact that this is not reflected in an increase in the strength of COD merits the serious consideration of Conference. It appears that the composition of members is homogeneous whereas the developing Nationalist opposition among the Europeans outside our ranks is from a diversity of opinion and outlook.
>
> There is an urgent need for our organisation to broaden its ranks and make every effort to establish contact and wherever possible cooperate with these people both inside and outside our organisation.[8]

These opposite positions begged a resolution. It arrived in the form of the COD President's speech. Father Jarrett-Kerr, C.R., the successor to Trevor Huddleston in Johannesburg, was invited to open COD's conference and read banned COD President Piet Beyleveld's speech. The members of Hillbrow might have viewed this invitation as foreshadowing the strategy the COD leadership would pursue. Piet Beyleveld entitled his address, "Where Do We Go From Here?" The apparent purpose of the speech was to address, squarely, criticisms of the organization: the success of government persecution in scaring off members and potential members; COD's inability to safeguard its members from persecution; the leadership's unwillingness to take COD underground; and the organization's unwillingness to work among the white working class and develop a dominant working-class ideology within itself.[9]

In response to these complaints, Beyleveld stated the leadership's unequivocal decision to organize on a broad front. By the later 1950s it was not uncommon to hear COD describe itself as "a loose association of like-minded people, bound together by a common belief in the necessity for a democratic society based on the equality of all citizens." Beyleveld said this but was even more explicit, suggesting that COD should be able to provide a home for "such diverse political allegiances as radical Black Sashes, Liberal Party members, Communists, non-party radicals and democrats of a dozen different ideologies and creeds." In its first years COD members had seemed honestly miffed at the organization's inability to attract members outside a small coterie of Communist Party members, non-party radicals and left-of-liberal progressives. By 1958 COD had been rebuffed many times, and COD leaders understood why the organization alienated so many individuals who were otherwise strongly anti-apartheid. Therefore Beyleveld's statement was a declaration that COD was going to attempt consciously to make itself more attractive to the politicized liberal white population at large:

> Admittedly this is C.O.D. as it should be even if it is not unfortunately C.O.D. as it is right now. But if that is the ideal for which we are striving, what can one say of the criticism that the concentration of our work should be amongst the European working class, and that our ideologies should be strengthened with socialist ideas? One can and must conclude that such criticism is ill-founded and wrong. It results from the attempt to turn C.O.D. from what it is -- a loose association of like-minded people -- into what it is not, a political party striving for state power, and bound by an ideology and discipline. Such attempts are, in fact, subversive of all C.O.D. is and has always attempted to be.[10]

Beyleveld, who had joined the underground South African Communist Party in 1956, was influenced by the climate fostering a broad front of Congress and liberal organizations who shared a minimum program of action with a view to national liberation and universal suffrage. Consequently, COD moved in that direction. Beyleveld chose to explain this modification of COD's program as a result of COD successes:

> That in putting forward our advanced and radical point of view without compromise we have not only awakened new thinking amongst the population generally, but we have moved every existing European organization -- Black Sash, Liberal Party, Labour Party and even sections of the U.P. -- to revise their former outlook and programmes and to move closer to the Congress concepts of equality. (Note for example the development of the Liberal Party policy on the question of the franchise, which has taken place against a background of our creating).[11]

Beyleveld took these apparent successes as "the vindication of the correctness of our views and of our policies" and chose to find nothing new in COD's revised orientation. Yet there was a seeming contradiction in his various statements. COD's "advanced and radical" point of view had won out over time, and the liberal doubters and antagonists had been brought around to COD's positions regardless of whether they chose to acknowledge the source. However, COD was now throwing open its arms to the liberals, and the president, a member of the SACP himself, was stating that, while individuals were welcome to subscribe to any ideologies and organizations they desired, there was no room for socialism as either an explicit or implicit ideology within COD.

Father Jarrett-Kerr put it succinctly during his address to the conference, stating that COD president Beyleveld's message seemed to him "to express admirably this common front point-of-view, and [I] can associate myself 100% with everything in it."[12] Jarrett-Kerr's "conversion" may have been a marked success for the Congress of Democrats', and the Congress Alliance's, new program of broad alliances, but for the Congress of Democrats it was not business as usual.

One of COD's dissidents was allowed to put his arguments in the pages of COD's internal publication *CounterAttack* following the conference. The thrust of "V.'s" argument was simple and straightforward: In hungering after alliances with liberals, COD had betrayed its own leading role among whites. COD had lost sight of its analysis when it refused to acknowledge its "leftness" and had turned away from the class-analytical approach that it had known to be correct and that was now part of the Freedom Charter's economic clauses:

> We do not represent ourselves as the "conscience of white South Africa" -- as a more successful organisation does -- or if we did perhaps it would be better if we were to dissolve COD and join this other organisation. Yet we try to show how "nice" we are, not only in an effort to convert the non-believer, but to form alliances with the Libs, Labs., Black Sash etc. Is it any wonder that these groups are suspicious of us?[13]

"V." asked that COD work among both white and black workers, and particularly the latter. African workers would play the decisive role in South Africa's future, and COD avoided them at the peril of future "bloody racial battles." "V." asked that COD return to the Congress, as "V." understood it, and acknowledge that its beliefs and policies set it apart from liberals. The Congress Alliance had to acknowledge the primary role of the African working class in the struggle and make plain its call for economic justice through nationalization of South Africa's natural and industrial resources. There was a clear strain of mockery in the editorial, insinuating that COD was in danger of becoming a poor and marginalized imitation of the Liberal Party.

"V." 's claims did not go unanswered. The official COD respondent to "V." described "V." 's arguments as sectarian and potentially capable of "destroying Congress altogether."[14] The respondent argued that the Congress Alliance was by nature a multi-class agglomeration, an inherent aspect of the partnership of the Congress affiliates, which would be abrogated were COD to follow "V." 's prescriptions.

How widespread was the sense among COD members that the organization had abandoned its principles and lost its way? How representative was "V." of opinion among the more radical active members of COD in the late 1950s? It is doubtful that "V." was highly representative. "V." appears to have lost sight of an important point in the rush to ideological purity: The policy of going "to the Libs. and Labs." did not originate with COD but was rather a function of the climate of the late 1950s, in which barriers between the Congresses and liberals had begun to break down. COD's drift to moderation was thus a function of COD's discipline as an acknowledged subordinate of the ANC. The ANC had been approached for the purposes of opening lines of communication with the broader white community and had accepted these offers. The initial contacts had expanded to include a broader section both of Congress activists and the white community. If the organizational tide had turned toward the creation of a broad united front, then COD's willingness to do its part in that direction was a function first and foremost of its relationship to the Alliance as a whole. COD's chartered purpose had been to reach into the white community as a representative of the ANC, SAIC, and SACPO. What had proved impossible earlier now became possible by virtue of the changes made in the political, economic, and social landscape by the Nationalists. "V.", by virtue of his or her point of view, was out of the "radical" mainstream in 1958.

COD and the Liberal Party: Building the United Front

In 1958 Professor Phillip Pistorius, one of the few Afrikaner intellectuals willing to take part in the multi-racial exercises of the time, engaged in a correspondence with the Liberal Party leader Alan Paton. Pistorius was afraid that if blacks could not be convinced that liberal thought would become a political factor of decisive importance, South Africa would face a "violent climax" to the then-current trends of separation and oppression. Pistorius noted his special uneasiness at the COD-ANC alliance. He considered COD to be in favor of "violent solutions and emotional upheavals."[15] Yet Pistorius noted that at a Liberal Party meeting which he addressed and where a number of COD Youth branch members were in the audience: "The question asked me by the Congress of Democrats [members] at that meeting, and to which I had no effective reply, was how I could resist all unconstitutional and violent methods or methods that could and presumably would lead to violence, such as strikes, and nevertheless hold out the hope that things will change."

Pistorius' personal crisis led him to enter the Progressive Party, a rare defection among Afrikaners, even Afrikaner intellectuals. Paton, in response, noted that although the words "boycott" and "strike" would have "distasteful associations" for Pistorius, it was difficult to know "what other ways are open to unenfranchised people."[16] Paton commented on Pistorius' struggle with the efficacy of extra-parliamentary action and the possibilities of unleashing violent tendencies:

> This same struggle that goes on in your own mind goes on in the Liberal Party also between those who hate violence so much that they even hate all kinds of pressure, and those who accept the necessity for these pressures, even though they concede that they may be dangerous. It seems to me that unless one accepts this second point of view, one might as well say, I was against apartheid and I still do not like it very much, but I am now prepared to work for it as the only hope for the future.[17]

The sentiments expressed in both letters were indicative of a certain political fluidity at the end of the 1950s. Pistorius was representative of that small minority of Afrikaner intellectuals who had begun to perceive the disparity between the ideal of separate development and the unfolding realities of apartheid. Paton's response encapsulated the Liberal Party's movement away from legalist thought. Paton clearly accepted the necessity of extra-parliamentary action. And by 1960 the Liberal Party had dropped its objections to an unqualified franchise. What Paton did not cast off was his distrust of the Congress of Democrats, a manifestation of his deeply held liberal beliefs, and his consequent distrust of any individual or organization he considered communist-influenced.

As noted above, the National Party's 1958 victory proved a decisive factor in moving Congress members and liberals toward a concord. Following the 1958 election, *Fighting Talk* published three articles on the question of unity among diverse anti-Nationalist organizations. These articles were sparked by a series of Liberal Party member Owen Vine's writings in the *Rand Daily Mail.* Vine had called for the creation of a working unity between all white anti-apartheid organizations, and his call had touched a nerve among many of those individuals involved in those organizations.

Father Jarrett-Kerr, Ben Turok of the Congress of Democrats, and Peter Rodda, then a Liberal Party member, were asked to comment on Vine's articles. Jarrett-Kerr suggested that some kind of immediate federation was possible, based upon whatever common ground the parties involved could find. The federation could then go to the ANC, SAIC, and SACPO and say, "Here we are: use us as you wish. We are dedicated to the achievement of justice and true democracy; but we are a minority -- you, the majority, must decide how we can best serve."[18] Turok spoke strongly in favor of Vine's idea. However he stated that groups other than his own, specifically the Liberal Party, had reservations about such a federation.[19]

Turok pointed to the commonalities among COD, the Liberal Party, the Black Sash, and similar organizations. While he did not call for amalgamation, Turok suggested that liberals would find they were progressively powerless without black assistance as Parliament lost all legitimacy. An obvious remedy was to link liberals to each other and then to Congress. Rodda agreed with much that Turok had to say, including the potential value of some type of common front and the impossibility of actual amalgamation at the time. Both Turok and Rodda conceptualized a multi-racial conference in the future as a means to link the proposed loose federation of white organizations with the Congress Alliance:

> I hope that a Consultative Committee will soon be set up, that this will lead to a loose Federation, which will later come under a common banner with the Non-White Congresses in a powerful Common Front. Probably this Front will

> emerge from a Multi-Racial Conference in the not too distant future if we all act with wisdom, vigour and integrity.[20]

There was significance in the fact that a debate between people who shortly before were mutually distrusting appeared to become an exercise in consensus-making. Was such unity realized in actuality? The answer varied from time to time and place to place but was, in the short-term, a qualified, fitful "yes."

COD's decision to run its own candidate in a Johannesburg by-election proved to be one of the first instances of COD-Liberal Party cooperation. COD had written to the Liberal and Labour parties soliciting support for their candidate. The Liberal Party, responding, suggested that the organizations support an independent candidate, Ruth Hayman, who had been a member of both COD and the Liberal Party but was then active in neither. The internal circular explaining the decision to withdraw the COD candidate was offered to the regions to allay suspicions or antagonisms about this electoral decision.[21] The only tactical comment offered on the decision is found in a letter from the COD secretary in Johannesburg, who stated that the decision was a correct one and would "be a strong lever in our favor and enable us to put pressure on the local liberals who are apathetic and anti-COD."[22] The COD secretary's comment suggests that while there was no love lost between COD and the Liberal Party, COD was going to work actively to implement the broad front and was not above calculation in moving the clearly reticent local Liberals into a closer alliance.

If COD's cooperation with liberals in the later 1950s was a matter of some controversy within COD, the relationship was more controversial still among liberals. Discussion of the relationship among leading members of the Liberal Party revealed a high degree of calculation and an ever-present fear of communist taint. In the wake of the Multi-Racial Conference the Liberal Party was attempting to walk a complex line between its principles, including a dedication to a free enterprise system in a liberated South Africa, and the acknowledged need to maintain an opening to the ANC through the Congress Alliance. The Liberal Party's relationship to the Alliance was made more complex by the fact that the issue of Liberal Party-African National Congress cooperation was ever present. The Liberal Party saw itself as a kind of third force between the Congress Alliance and the other, segregated, liberal opposition organizations. It considered its non-racial structure to be a key aspect of its relevance on the political scene. Yet the Liberal Party was never prepared in the 1950s to accept the hegemony of the African National Congress and to place itself in a formally linked and subordinate position. Thus, the Liberal Party remained a competitor, and suspect, in the eyes of the leadership of the African National Congress. The Liberal Party appeared to be in a perpetually unstable orbit around the ANC, a body by which it was both attracted and repelled.

If the Liberal Party found it impossible to resolve its relationship to the ANC, there were Liberal Party members who had clearly resolved to complicate the Liberal Party-COD relationship whenever possible. Patrick Duncan was a maverick member of the Liberal Party leadership who had taken part in the Defiance Campaign and would later join the Pan-Africanist Congress (PAC). Duncan's relationship with COD was indicative of the difficulties that complicated any Liberal Party attempt to work with the Congress of Democrats. Duncan was the editor of *Contact*, which was more Duncan's personal mouthpiece than an official Liberal Party newspaper. However, the paper did publish a regular column by Alan Paton and Peter Brown, two leading Liberal Party leaders.[23] In late 1958 Duncan, as editor of *Contact*, refused to publish a COD advertisement on the grounds that COD was a "totalitarian organisation."[24] The COD correspondence noting this refusal also remarked that Duncan had made this decision without first referring it to his editorial board. Duncan was given to eccentricity and exhibited a tendency to put his personal agenda before that of his party. Alan Paton, then President of the Liberal Party, was much more sensitive to the potential damage incurred by Duncan's outbursts:

> Pat has put us in another crisis: he refused to place a COD advert in CONTACT, advertising a Treason Trial meeting. He completely fails to see that this strikes at the very heart of the Multiracial Conference, and at the Party policy of cooperation on agreed objectives. What will he say at Accra?[25] Launch a full-scale attack on Communists in Africa, and on all African National Congresses for being duped by them? It could easily happen, because Jordan [Ngubane] is also inclined that way, and these two could quite easily bring the Party into bad odour. It is already being said that Pat is really pro-Nationalist! If we broke our tenuous relationship with the COD now, the ANC and NIC and SAIC would slam their doors in our faces.[26]

Paton's statement suggests two important elements of Liberal Party thinking at the time. The Liberal Party had valued the Multi-Racial Conference as a means of establishing closer ties with the Congresses. The Liberal leadership was willing to swallow some of its pride and police its membership if that was a necessary aspect of the unfolding ANC-Liberal Party relationship. Paton's fear, expressed at the end of the correspondence, indicated that the Liberals' "tenuous" relationship with COD was also a strictly instrumental relationship in Liberal minds. Taken at his word, Paton appeared to admit that mutual activity with COD was a means of establishing ties with the ANC. COD was to be suffered for this purpose. COD chose not to publicize the episode with Duncan.[27]

In fact, a note from the Cape Town COD in October 1958 advised that the region was to hold a meeting, which Duncan would address. COD had recently held a meeting about Parliament that members of the Liberal Party and Black

Sash had attended. Minutes of COD National Executive Committee meetings in early 1959 evidence interest in joining COD-liberal efforts. The COD Secretariat had discussed the possibility of closer cooperation with the Liberal Party. The Secretariat had considered the time right to push for the creation of a permanent liaison committee between the two organizations.[28] Minutes of a March meeting of the COD Executive cited reports that the Liberal Party had indicated an interest in gaining representation on the Congress Consultative Committee. The Executive did not comment on this particular possibility but was clearly in favor of moves toward unity between itself and the various liberal organizations:

> It was also reported that there were moves for liaison between the Federal Party, Liberal Party, Black Sash, Labour Party on the initiative of the Federal Party. We are watching the position closely. It was felt that we have little to fear as none of these organizations can be really effective without cooperation with the Congress movement. It was decided that we should urge that the committee which came into being on the mixed gatherings ban and on which all organisations are represented, should be activised [sic].[29]

For the Liberal Party leadership, contact with the Congress Alliance and the ANC were the prizes to be won from these contacts. COD's vantage point was quite different. The leadership of COD considered the organization to be in a prime position as the chief white ally to the Congresses. COD's only conceivable fear was that of a Liberal Party-ANC liaison taking place that might lead to COD's loss of position in the forefront of the liberation movement, a fear that was unlikely to be realized. On 14 July, 1959, the Transvaal Provincial Committee of the Liberal Party recommended that the organization establish an informal liaison with COD.[30] Two months later the same committee again took up the question of cooperation. Discussing an upcoming election, the Liberal Party's Provincial Committee sought to whip up electoral support for its candidate among COD members while counseling caution in accepting COD's aid: "Members of COD [are] not to be asked to canvass but if any offer, this [is] to be accepted but the [COD] person must be accompanied by a member."[31]

Liberal Party members distrusted their counterparts in the Congress of Democrats while acknowledging that COD was their primary competition for a good working relation with the other Congresses and the black population. Writing to Peter Brown, Jack Unterhalter, also a leading Liberal Party member, recounted how the Bishop of Johannesburg had mentioned to another party member that the only whites who had helped the Frenchdale detainees with food and clothing were "COD types." Unterhalter asked Brown if the party could get involved also.[32] Yet Liberal Party members would point to COD members' seemingly effortless movements into and out of the townships as if

Liberals faced an impenetrable barrier that kept them from maintaining a presence there:

> Several very influential Africans have joined the [Liberal] Party in the East Rand and this is where membership is expanding most rapidly. Many people are willing to hold house-meetings in Kwa Thema and other Springs townships, but there aren't any speakers and Europeans cannot get permits to enter the townships. Meanwhile, COD people just do enter and do their best to dissuade one and all from joining us. To try to solve this, Julius will spend every other weekend out there.[33]

The barrier that kept the Liberal Party out of the townships was largely of its own making, fabricated of Liberal fears, a lingering legalism, the desire to make political involvement as clean as possible (rather than skulking around African areas on the periphery of the law), and an abiding distrust and objectification of "COD types" that kept Liberals from engaging in activities that appeared redolent of such radical taint. The Liberal Party leadership sensed that, holding themselves at arm's length from the Congress movement, from 1958 they were standing on the periphery at a critical moment in their own history, threatened by the more moderate Progressive Party. There is no evidence that COD members consciously attempted to dissuade African participation in the Liberal Party. It is more likely that COD members relied on the Liberals' own limitations -- those self-imposed barriers -- to moderate African interest in the organization.

The Liberals' self-stated dilemma around entering the townships is an example of the nature of the divide separating radicals and liberals. Bringing material aid to the African Frenchdale detainees was probably as natural an act to COD members (who often brought things to their own members in prison) as it appeared daunting to Liberal Party members. COD members accepted that they would have to break apartheid laws to carry on the necessary work of liberation. Yet, in 1960, liberals continued to fret about obtaining police permits for township entry. In October 1959 Peter Brown complained to Jack Unterhalter that the Progressive Party was portraying the Liberal Party as slightly dangerous in order to encourage anti-apartheid whites to take up the Progressive Party as a moderate alternative. Brown conceived that under this type of assault the Liberal Party could be delegitimized and "driven toward the COD position where we really have no influence over white opinion whatsoever."[34] In Liberal minds, any attempt to build up a black constituency was fraught with danger, and would antagonize the ANC. Yet the party leadership fought their perceived peripheralization. The Transvaal's monthly report of July 1960 noted that an informal plan had been suggested in Johannesburg whereby party members would drop by the various offices of the Congress affiliates and "casually exchange views": "We are doing the same thing with the COD and hope, in this way, to know better about what is going on."[35] This was the liberal paradox: To "know what was going on" one had to take risks and shed concern for one's image in the context of parliamentary politics. The Liberal Party wanted to share in the benefits of being close to those in the know while keeping them at arm's length for appearance' sake. This was ultimately an untenable position.

The "Common Front" experiment was a failure, and understandably so. How could the white liberal community hope to link up with the Congress Alliance while its constituent members were jockeying for "holier than thou" positions amongst themselves? The years between the run-up to the Multi-Racial Conference and the end of the decade offered some possibility of wider unity. The prospective players were driven by the realization that the Nationalists were in power to stay, the United Party was a dead letter, and, barring all-out war, the only hope for a political solution lay in strength in numbers and a common, minimum program that might sway the white community from *baaskaap.* However, this program was never defined in concrete terms. The participants aided one another in piecemeal ways, met irregularly and maintained their respective sovereignties. The initiatives to unity had been taken by liberal members of the white political and religious communities, and the ANC had happily embraced these efforts. However, in the end, the ANC could not save white liberal politics from its own contradictions. If COD was not the ideal white partner organization, and indeed there were tensions between the ANC and COD in these years as discussed below, the ANC was apparently satisfied that COD was largely a loyal and hard-working friend. The ANC could welcome the Liberal Party and the Progressive Party, and even work with them, but could it trust them at the end of the day? The answer was "no."

COD, the ANC, and the "One Congress" Question

The Liberal Party had considered one of its strongest drawing cards to be its non-racial membership policy. The party used this fact rhetorically to bludgeon the Congress of Democrats and the Congress Alliance, pointing to their racially separate organizational schema. An article in *Contact* noted that the Transvaal Indian Youth Congress had proposed a resolution calling for the amalgamation of the Youth Congresses into a single organization, which the *Contact* writer considered a "tribute to the non-racial basis of the Liberal Party, which is still the only purely non-racial body in the country, and which owes its recent striking successes to its principled adherence to this basis."[36] The organizational structure of the Congress Alliance was a sensitive issue, particularly between the ANC and COD. There had been rumblings within COD about the amalgamation of the Congresses since COD's formation. COD members, speaking some years after the fact, had a tendency to dismiss these challenges to the Congress Alliance from within as inconsequential, the product of personal eccentricities or youthful enthusiasm. Yet by the late 1950s both the COD leadership and the ANC were sufficiently concerned about these "rumblings" to issue internal statements and to use public opportunities to defend against charges of Congress racialism while defending the Alliance as it was. The "One Congress" question was particularly sensitive, given the charges over the years that the Congress of Democrats was attempting to "lead the ANC by the nose."[37]

Those publicly raising the One Congress question were not necessarily members of COD. Ronald Segal, the editor of the journal *Africa South* and a friend to the Alliance though never a member, attended a delegate conference of the Congress affiliates in Johannesburg in 1958. There Segal likened the organization of the Alliance to the Group Areas Act. He noted that ANC leaders had stated before that they would open the organization to other races if they had the assent of ordinary ANC members. Segal took a pound out of his pocket as eight years' membership dues and asked the audience, "will you take me?"[38] Segal continued his crusade to open up Congress in the pages of *Fighting Talk*. In an editorial calling upon the ANC to " 'unlock the gates' " Segal suggested that whites had never been happy with the organizational form of the Congress Alliance, that Indians' anxieties were increased by their separation from Africans, and that Coloureds required "the concrete re-assurance of political work with Indian and African" before they could surrender their fears. Segal asked if the ANC was not unwittingly "playing the game of the Africanists?"[39] This was not the only such instance. Two socialists, renegade COD members Baruch Hirson and Vic Goldberg, were brought before an *ad hoc* COD disciplinary committee for the same offense:

> It was reported that the regional committee has referred the question of the two members who made public statements about wishing to join the ANC to the NEC. The NEC thereupon discussed the matter on the basis of the following facts: Mr. Hirson and Mr. Goldberg made these statements at a delegate conference of the Congresses. Mr. Goldberg had in fact been sent as a delegate by a joint branch meeting, and was wearing a Congress uniform at the Congress. Mr. Hirson was not sent as a delegate, but sat on the platform, and was probably taken to be a COD rep. at the meeting. A letter from Mr. Goldberg was read to the NEC in which he stated his political position in relation to COD. After some discussion the matter was left in abeyance till the next meeting.[40]

In a subsequent meeting of the NEC, it was noted that Goldberg had apologized for his actions, and had stated that rumors to the effect that he was planning to travel to a meeting of the ANC Youth League in Durban to call for a unified Youth Congress were "quite unfounded."[41] However, after the reprimand and his apology, Goldberg continued to call for a single Congress organization. He wrote a short rebuttal piece to an article in *CounterAttack*, entitled "One Congress," some time after May 1959. Goldberg pointed to Ronald Segal's experience, when the largely African crowd had appeared to approve his bid for ANC membership, as proof that the masses were ready for a non-racial Congress. Goldberg suggested that COD was primarily concerned with white interests when in fact COD members were concerned with oppression generally and the breaking down of racial and ethnic barriers.[42] His paper elicited responses from both COD and ANC spokespeople. The COD respondent charged with writing a rebuttal to Goldberg appeared rather to bolster Goldberg's argument, acknowledging that most members of the Congress of Democrats would prefer to work in a unified Congress organization:

> Such an organisation would be more in keeping with the type of society we are working for, and we would also be able to experience more directly the stimulation that contact with our non-white fellow congressmen gives. The feeling of strength aroused by 1,000 voices singing *Nkosi Sikeleli Africa* and the heights of solidarity and heroism that the African people have shown under pressure has bound many a white democrat closely to the movement for liberation....In general, such experiences serve to confirm and reinforce the theories and abstract values that bring whites into the struggle.[43]

This was a serious admission because it belied the notion that the "One Congress" sentiment was confined to individuals outside COD or fringe elements within. The personal relationships that had grown between black and white members of the Alliance were an important aspect of the Alliance's work. Still, the COD writer agreed that Africans had "the right and even duty" to form an organization of their own to defend themselves. The problems of South Africa's communities differed, and if a single Congress were formed, new African, Indian, and Coloured national organizations would undoubtedly be formed to meet the needs that the Alliance partners had met. The writer warned that COD members had much work to do among the white population, and that work would become more important as the Nationalists continued to stir up white fears. The writer also warned that COD members would be ill-advised to "urge a change the whole Congress leadership has now expressly opposed." The ANC's response to Goldberg's statement was similar to that of the COD writer. The ANC writer suggested that the assembled would not have turned down Ronald Segal's request to join the African National Congress for fear of appearing to refuse cooperation. However, "there is absolutely no doubt that had these same delegates been presented with the problem in a debate they would not have accepted the suggestion."[44] In fact, the writer noted that the whole of the ANC leadership considered the call for a non-racial Congress incorrect.

Most COD members and other progressive whites would have preferred to work in a non-racial Congress. However, they were politically disciplined individuals who understood their subordinate position to the ANC. It was the ANC's perception that the objective conditions did not warrant, and that their membership did not desire, the amalgamation of the Congress affiliates. Hirson, Goldberg, and Segal were vocal exponents of their cause. Their potential for damaging relations between the Congresses may have been reduced somewhat by virtue of their fringe relationships to COD.[45]

COD's most serious challenge to the Alliance status quo on this question came from the organization's youth members, particularly its university groups. The question of the place of youth in COD and the relationship of youth to the "One Congress" question was taken up at a COD National Council meeting in 1959. There were differences of opinion among COD leaders on the question of whether there should be separate youth branches or whether youth should be members of ordinary branches. Youth members had complained that ordinary branch meetings were "insufficiently vital."[46] At the time of the meeting a non-racial pan-Congress youth group was in operation at the University of Cape Town. In April 1959, the COD Youth branch at the University of the Witwatersrand had voted to disaffiliate from COD.[47] A Johannesburg Youth representative, speaking to the National Council, explained that there was a need for a separate Youth branch because there were a number of students at Wits who thought of themselves as Congress rather than COD supporters:

> One of our goals was the achievement of a multi-racial organisation. There are many obstacles towards achieving this. The Liberal Party and the former CP were multi-racial. For one thing, care has to be taken that one group does not dominate the other. The ANC Youth League was not nearly as strong as it used to be. One reason for this was that the ANC had been dampening down the ANC Youth League. Since the advent of the Pan-Africanist movement, a strong effort had been made to seek the support of young Africans. A multi-racial Congress where white and non-white youth found themselves together would counteract this. If a start were made, non-white youth might not display such unwillingness to enter an inter-racial organisation.[48]

The chairman of the National Council noted that the Youth branch at Wits had been criticized for attempting to change the organization of Congress, and the other Congresses had also complained about the Youth branch's activities. One of the Youth representatives agreed that the COD disaffiliation had "aroused immediate and evident antagonism in the other Congress youth organisations." The decision was therefore not implemented. The chairman acknowledged that COD members had never questioned the idea of a single Congress "in principle," but there were many obstacles in the way of the realization of a single Congress body. A committee of the Youth branch recommended that it remain affiliated to COD but be open to youth irrespective of race. The Youth branch report did indicate that the youth hoped a unified Youth Congress would soon be formed, to function "as the historical origin of a unified Congress." In closing discussion of the place of the Youth branches and the question of pan-Congress Youth branches at the universities, the chairman stated that there was merit in the establishment of such "Congress youth groups," and this would be discussed with the other Congresses in the National Consultative Council. However, the chairman did state that the youth should wait until the ANC Youth League took the initiative of proposing such an amalgamation.

Piet Beyleveld had raised the "One Congress" question in his address written for the National Conference in 1958. Beyleveld admitted that there were people who chose to stand outside of COD only because of the racially based organizational form of the Alliance. But he was insistent that the time for a single Congress had not arrived:

> It will have to be fought for; all the old prejudice of race against race will first have to be broken down; all the deep ingrained racial feelings of inferiority and superiority will have to be wiped away; all the residential and group areas bars to real multi-racial cooperation will have to disappear. Until that time, we believe that multi-racial organisations will be capable at best only of organising the most advanced, independent and conscious people of all races -- not the masses.[49]

Beyleveld's response in defense of the Alliance reflected the ANC's own thinking closely. The "One Congress" debate served to highlight the gulf of understanding that persisted between white progressives and the other members of the Alliance.

What was Peter Rodda, a sometime member of both COD and the Liberal Party, thinking when he wrote that the "historic reasons for this [Congress] structure have clearly disappeared"?[50] What were Rodda's referents in making that determination? That blacks and whites appeared to work together in a comradely and mutually trusting atmosphere in the Liberal Party? That the Congress leaderships appeared to share a strong respect for and trust in one another, and a vision of a non-racial future? This type of thinking, which the proponents of "One Congress" held in common, mistook *one* aspect of Congress strategy and philosophy for the central and moving force behind the ANC and the Congress Alliance. The ANC had never pretended that it was other than a nationalist organization, working for the achievement of the African people's freedom. That had always been, and would continue to be, its primary goal. Yes, the ANC had worked to create alliances with the Indian community and to create allied organizations among the white and the Coloured communities. Yes, the Freedom Charter had amplified the vision of a South African nationality that transcended racial and ethnic divisions. And if whites responded particularly to the non-racial aspects of Congress work, that is not surprising. Coming from positions of relative material comfort, it was to be expected that whites who identified with the liberation movement would hope to see Congress policies modified in a manner over time that made them feel welcomed and as integral to the national liberation movement as possible. However, in their preoccupation, whites lost sight of the central question, which was a question of power. Two of the ANC's central goals were: the restitution to the African people of their rightful place as politically free and materially secure citizens in a free South Africa; and the creation of a new South African identity without reference to race or ethnicity. The ANC did not weight these goals equally in 1959. African liberation was first and foremost on the agenda, and the bottom line was the empowerment of the powerless. In its report to the Annual Conference in 1959, the ANC's National Executive described the organization as under attack from two directions: The Africanists were attacking the ANC for its alliances outside African circles, and other, supposedly sympathetic forces were attacking the ANC for being narrow and tacitly accepting racialism.[51]

Given this perceived double barrage, the ANC National Executive felt it was necessary:

> Once and for all to get the record straight. Neither the ANC -- or for that matter any of the other Congresses -- were formed for, or exist for the primary purpose of building a "multi-racial" or "non-racial" society. The ANC was formed to unite and voice the views of Africans. That remains its primary purpose. Let those who will, call this "racialism". But most people who look at our achievements honestly and without malice will realise that the building of an all-Union organisation of Africans, built in the teeth of every obstacle that governments could muster against it, and the leading of that organisation to become a mighty power in the land is an achievement from which not only Africans but all democratic South Africans can draw pride and satisfaction.[52]

Duma Nokwe, writing for the National Executive of the ANC, described Africans as subject to special laws and a unique level of oppression, which dictated that the grievances, aims, and outlook of Africans were not identical with those of Indians, Coloureds, or whites: "If tomorrow there were no African National Congress, we would have to set out again, from the beginning, and build one." The language of this statement, which was later reprinted in full in *Fighting Talk*,[53] was strong and straightforward, intimating that the ANC was not in the mood to brook further discussion on the issue. Nokwe pointed to "European supporters of the democratic cause" as the principal instigators of the controversy. Describing COD as "an organisation of non-conformists from the ranks of the oppressor caste," the report agreed that while the interests of the ANC and COD might converge in the long term, their interests were different in the present. That Nokwe framed COD in the context of the "oppressor caste," a rare and pointed reference, suggests the frustration, and possibly the bitterness, the issue had engendered. Whites were cautioned not to "seek to 'reform' the Congress movement to suit their special and unique position." Nokwe praised the Alliance "as the high water mark of racial fraternity and cooperation," having raised a new generation of South Africans "in the mould of equality and brotherhood." The ANC decided by mid-1959 that there had been sufficient discussion around the "One Congress" question. Without wanting to demoralize or embarrass their white supporters, it was necessary to make its white allies understand that no matter the level of unity achieved among the leaderships, the African in the street was not prepared to cede his one structure of self-determination, the ANC, to non-Africans:

> After all, everything that can be done in this country is done to make the African feel and appear to be helpless and inferior. There is no aspect of his life where he can act for himself without White supervisors, superintendents, foremen and bosses. What could be more natural, then, that at least in the African National Congress members wish to make certain that this is truly their own organisation?
>
> It is embarrassing to have to point this out to our friends, and we should like to make it clear that we do consider them friends, although we disagree on this matter; we respect their motives. All the same, they have thrust this embarrassing discussion upon the movement.[54]

The raising of the "One Congress" question was undoubtedly embarrassing not just to the ANC but to the COD leadership, who found themselves answering for their members as well as renegades and individuals who had no affiliation to COD. Their position may have been aggravated further by the fact that, if they were not vocal exponents of the "One Congress" position, they were quite likely to be silent adherents. COD leaders were also faced with the challenge of answering the Liberal Party's charges of racialism-by-segmentation. Those who raised the question may have come to the realization that in attempting to bring the Congresses closer together they had ignited one of the most controversial episodes in the short history of the Congress Alliance. The raising of the question reflected a kind of social and political blind spot that was unique to whites, who the ANC spokesperson above described as having a "special and unique position" in the national liberation movement; this was coded language suggesting an entrenched, if narrowly focused, difference of outlook. For the majority of whites (albeit a minority of COD members) who pushed for the creation of a unitary Congress, the driving desire was less a matter of principle than an inability to see the Congress Alliance for what it was: a significant, even unique, alliance of organizations who shared a set of broad and common goals and which had been formed at the behest of the ANC in the interest primarily of tactical advantage in the fight against the implementation of apartheid. Whites experienced day-to-day relationships with blacks through the Alliance, and they may have considered this to be the most valuable product of the Alliance. Africans did not. For the leaders of the ANC, the very fact that the "One Congress" question was raised may have provided the strongest evidence that misunderstandings and insensitivities between the Congresses persisted and that the time for a unitary Congress had not arrived.

A Pan-African Postscript

The "One Congress" controversy is illustrative of the tensions arising from attempts to build a coherent alliance across racial and ethnic lines. Yet, COD members' belief in the uniqueness and importance of the Alliance's multi-racial form and non-racial atmosphere led COD to seek, of its own volition, recognition for the Alliance and COD's own unique position therein, in Pan-Africanist circles outside South Africa's borders.

In late 1958 the Congress of Democrats made representation to the first Pan-Africanist Conference held in Accra, Ghana. Identifying themselves as "white victims of the mad and cruel policy known as 'white supremacy,' " the members of COD introduced themselves as representatives of a "more far-seeing" minority.[55] The COD memo attempted to explain that South Africans faced something more complex than a simple clash of black and white:

> White South Africans have become, over the generations, part of the settled permanent population of Africa, knowing and claiming no homeland outside of Africa. For white South Africans generally, there can be no prospect of retreat under pressure from Africa to metropolitan lands, as there may be for the white colonials of Uganda or the Dutch of Indonesia. They are now a part of the indigenous peoples of Africa -- a new people drawn from Dutch, French, British and German immigrants of long ago.[56]

The COD memo described COD's formation and position in the Congress Alliance, and the sacrifices COD members had shared with their black colleagues. An effort was made to stress the humility of COD members in light of the fact that when liberation dawned, "white South Africans [would] be a minority group not only by virtue of their numbers, but also by reason of the political power and influence those numbers will entitle them to wield." In COD's final plea, the emerging independent states of Africa were asked to proclaim that African liberation did not equate with "the cause of white extermination":

> Give new hope and courage to all in Africa who -- like us -- struggle for the future in the midst of multi-racial countries, by proclaiming that OUR future, the future of liberated Africa, lies in multi-racial societies, in which every

> minority people will be entitled as of right to all the liberties and privileges of the citizens of a democratic state.[57]

This was not COD's sole attempt to identify or affiliate itself with a Pan-African organization. In early 1960 Ben Turok wrote to the All-African Peoples Conference, asking for information and inquiring as to the possibility of affiliation. Turok enclosed a copy of the memo to the Pan African Conference.[58] Why these efforts to make Pan-African organizations aware of COD and the multi-racial nature of the Congress Alliance? The answer relates to a sense of South African uniqueness held by some white activists and the desire to see the microcosmic Congress example of multi-racialism and its vision of a non-racial future carried into other parts of Africa:

> These Pan African movements tended to be rather narrow, African nationalist, and we were aware of that, and some people in the movement thought that we ought to try and spread the sort of Congress Alliance idea widely into Africa. Because the small number of contacts we had indicated that black leaders, African leaders of the rest of Africa did not have our perspective, and so we began to put ourselves on the map. We did apply for affiliation to various things.[59]

COD's petitioning illuminates an important aspect of the organization and of the Congress Alliance. COD was a unique organization in an alliance unique to Africa. For those involved, this participation provided its own inspiration. Alliance members were traveling a difficult and dangerous political and social path. COD members were particularly conscious of the uniqueness of their endeavor, to the extent that they considered it a kind of socio-political experiment, with ramifications for the whole of Africa. The push to affiliate with Pan Africanism was thus a kind of odd missionary activity, with a base in liberation rather than in colonization. COD members, as the recipients of the beneficence of their Alliance partners, sought to share the possibilities of non-racial cooperation with other fledgling African nations.

The last years of the 1950s were insecure years for all those fighting the National Party's hegemony. As the government's legislative and coercive program accelerated, liberals found African nationalist initiatives and multi-racial cooperation more palatable, and the ANC and its partners, unsure of how best to proceed in the face of the Treason Trial arrests and generally repressive atmosphere, found the possibility of alliances with liberal and religious organizations a sane tactical strategy. The Congress of Democrats, its key leadership depleted in these years, suffered from a lack both of intellectual and organizational vigor. COD members generally attempted to embrace the united front concept and extended a hand to the Liberal Party. Internally, COD members, both those willing to follow the ANC's lead in building the united front and those antagonistic to it, sought to make sense of the changed circumstances in which the national liberation movement was operating. Those COD members who worked with the Liberal Party faced that organization's distrust. The last years of the decade were far from satisfying for those seeking a visible sign that apartheid's days were numbered.

One of the few constants for COD members in this period was the prized relationship with the ANC and the other Alliance affiliates. This link, on both personal and organizational levels, was palpable, a constant reminder of the possibilities of breaking through and building a democratic, non-racial society. As suggested by the "One Congress" controversy, the ANC had a keen understanding of what this relationship meant to its white counterparts, and African National Congress leaders themselves could appreciate the special nature of their relationship with COD members, which had been tested under conditions of duress. Still, the limitations of those relationships in the face of primary goals -- the end of apartheid and justice for Africans -- were manifest. COD and the ANC ended the decade having shared a rich set of experiences that would mark each of them and inform their future thinking. But in the context of apartheid society, Africans and whites remained separate, and, the Freedom Charter notwithstanding, race was still very much a live variable in the Congress equation.

Notes

1. Janet Robertson, *Liberalism in South Africa, 1948-1963* (Oxford: Clarendon Press, 1971), 196-197.
2. Correspondence, Chief Lutuli to Prime Minister Strijdom, 28 May 1957 in Karis, Carter and Gerhart, Vol. 3, 397-403.
3. "South Africa's Multi-Racial Conference: A New Approach to Race Relations," booklet, issued by the Planning Committee of the Conference, n.d., Mary Benson, personal papers.
4. Ibid.
5. H.J. Simons, "Nothing New from SABRA," *Fighting Talk,* Vol. 12, No. 5, August 1958.
6. Ibid.
7. "South Africa's Multi-Racial Conference: A New Approach to Race Relations," booklet, Planning Committee, n.d.
8. Congress of Democrats, 1958 Annual National Conference, "Resolutions to National Conference," Bellevue branch, 9 June 1958, SAIRR Political Documents, Microfilm 860, Institute of Commonwealth Studies.
9. Congress of Democrats, Annual National Conference, 1958, President's Address, "Where Do We Go From Here?," COD Microfilm, CAMP Microfilm 1671, Section 13, 1958 National Conference.
10. Ibid.
11. Ibid.
12. Congress of Democrats, 1958 National Conference, minutes, COD Microfilm,

CAMP Microfilm 1671, Section 13, National Conference.

13."After Congress," "V.," n.d., COD Microfilm, CAMP Microfilm 1671, Section 13, 1958 National Conference.

14. "After Congress -- Another Viewpoint," "R.A.L.," n.d., COD Microfilm, CAMP Microfilm 1671, Section 13, 1958 National Conference.

15. Correspondence, Professor Phillip Pistorius to Alan Paton, 24 October 1958, Alan Paton Papers, Microfilm 865, Institute of Commonwealth Studies.

16. Correspondence, Alan Paton to Professor Pistorius, 7 November 1958, Alan Paton Papers, Microfilm 865, Institute of Commonwealth Studies.

17. Ibid. As to COD, Paton wrote: "I have given much thought to the alliance of the ANC and COD, but I simply dare not start a discussion on this. It is so complicated that it will have to wait until we meet."

18. Father Martin Jarrett-Kerr, C.R., "Three Levels of Cooperation," *Fighting Talk*, November 1958.

19. Ben Turok, M.P.C., "Unity in Action," *Fighting Talk*, November 1958.

20. Peter Rodda, "Thoughts...in a Dry Season," *Fighting Talk*, December 1958.

21. Congress of Democrats, "Circular to all regions," 1956, COD Microfilm, CAMP Microfilm 1671, Section One, Minutes and Correspondence.

22. Correspondence, Y. Barenblatt to COD, Cape Town, "Letter C.T. 9," 6 June 1956, COD Microfilm, CAMP Microfilm 1671, Section One, Minutes and Correspondence.

23. Douglas Irvine, "The Liberal Party, 1953-1968" in Jeffrey Butler, Richard Elphick, and David Welsh, eds., *Democratic Liberalism in South Africa: Its History and Prospect* (Middletown, CT: Wesleyan Univ. Press, 1987), 130.

24. Congress of Democrats, "Sub-Committee Reports, Cape Town," 16 December 1958, SAIRR Political Documents, Microfilm 860, Institute of Commonwealth Studies.

25. "Accra" is a reference to the first Pan-African Congress at Accra, Ghana in 1958, which Duncan attended.

26. Correspondence, Alan Paton to Leslie, 25 November 1958, Alan Paton Papers, Microfilm 865, Institute of Commonwealth Studies. "Leslie" is most probably Leslie Rubin, a leading member of the Liberal Party and for some time a Natives Representative in the Senate along with William Ballinger.

27. Congress of Democrats, National Executive Committee minutes, 16 December 1958, SAIRR Political Documents, Microfilm 860, Institute of Commonwealth Studies. The minute stated that when approached by COD, *Contact* had agreed to publish the advertisement, and the NEC had acquiesced to the Liberal Party's request that the matter not be publicized because COD considered the matter to be between themselves and Duncan, rather than the Liberal Party proper.

28. Congress of Democrats, Minutes, National Executive Committee, 20 January 1959, SAIRR Political Documents, Microfilm 860, Institute of Commonwealth Studies.

29. Congress of Democrats, Minutes, National Executive Committee, 3 February 1959, SAIRR Political Documents, Microfilm 860, Institute of Commonwealth Studies.

30. Liberal Party, Transvaal Provincial Committee, Minutes, 14 July 1959, Liberal Party Papers, Transvaal, Microfilm 837, Institute of Commonwealth Studies.

31. Liberal Party, Transvaal Provincial Committee, Minutes, 10 September 1959, Liberal Party Papers, Transvaal, Microfilm 837, Institute of Commonwealth Studies.

32. Correspondence, "Peter" to Jack Unterhalter, 22 October 1959, Liberal Party

Papers, Transvaal, Microfilm 837, Institute of Commonwealth Studies.

33. Liberal Party, Transvaal, Monthly Report, August 1960, Liberal Party Papers, Transvaal, Microfilm 837, Institute of Commonwealth Studies.

34. Correspondence, "Peter" to Jack Unterhalter, 22 October 1959, Liberal Party Papers, Transvaal, Microfilm 837, Institute of Commonwealth Studies.

35. Liberal Party, Transvaal region, Monthly Report, July 1960, Liberal Party Papers, Transvaal, Microfilm 837, Institute of Commonwealth Studies.

36. "Unite into One Movement," *Contact,* Vol. 2, No. 10, 16 May 1959.

37. The speaker mentions this and other allegations in an address to COD's Annual National Conference, "Where Do We Go From Here?," 1958.

38. Ronald Segal, *Into Exile* (London: Jonathan Cape, 1963), 210-212.

39. Ronald Segal, "The A.N.C. Should Unlock the Gates," *Fighting Talk*, July 1959.

40. Congress of Democrats, National Executive Committee Minutes, "Transvaal," 9 February 1959, SAIRR, Political Documents, Microfilm 860, Institute of Commonwealth Studies.

41. Ibid., 30 June 1959, SAIRR, Political Documents, Microfilm 860, Institute of Commonwealth Studies.

42. This seems to have been a willful and gross misreading of certain statements made by COD for public purposes. Certainly some of COD's appeals to whites were couched in the language of self-interest, but any white familiar with COD's work was well aware that that was a simple rhetorical position.

43. "Supplement in Defence of the Congress Alliance," n.d., n.a., COD Microfilm, CAMP Microfilm 1671, Section 6, Supplement.

44. "An ANC Member's View," n.d., n.a. COD Microfilm, CAMP Microfilm 1671, Section 6, Supplement.

45. Goldberg and Hirson were socialists who entered COD only in the late 1950s, and were not allowed to return as members following the Emergency in 1960. Hirson was involved in the production of a number of publications highly critical of Congress while a member of COD, including *Analysis* and *Lekhotla la Basebenzi*. Segal was not a member of any political organization at the time, although he aided various Congresses on an *ad hoc* basis throughout the 1950s and early 1960s, and printed articles by Congress members in his journal, *Africa South*, and later *Africa South in Exile*.

46. Congress of Democrats, Minutes, National Council Meeting, 1959, SAIRR, Political Documents, Microfilm 860, Institute of Commonwealth Studies.

47. Congress of Democrats, "Youth Branch Report to National Council," 24 May 1959, SAIRR, Political Documents, Microfilm 860, Institute of Commonwealth Studies.

48. Ibid.

49. "Where Do We Go From Here?", Congress of Democrats, 1958 National Conference.

50. Peter Rodda, "The Africanists Cut Loose," *Africa South*, July-September 1959.

51. African National Congress, "Executive Report, 1959 Durban," Albert Lutuli Papers, Microfilm 845, Institute of Commonwealth Studies.

52. Ibid.

53. Duma Nokwe, "The High-Water Mark of Race Co-operation Is In the Congress Alliance," *Fighting Talk*, July 1959.

54. "Fusing the Congresses?", n.a., *Liberation*, No. 37, July 1959.

55. Congress of Democrats, Memorandum to the Pan African Conference, Accra, 5-12

December 1958, Signed by Piet Beyleveld, President, and Ben Turok. COD Microfilm, CAMP Microfilm 1671, Section 11, Leaflets.

56. Ibid.

57. Ibid.

58. Correspondence, Congress of Democrats, to the Administrative Secretary, All-African Peoples Conference, Accra, Ghana, 4 February 1960. Signed Ben Turok, M.P.C., National Secretary, COD, SAIRR, Political Documents, Microfilm 860, Institute of Commonwealth Studies.

59. Interview, Ben Turok, London, 27 January 1987.

10

White South Africans and Armed Struggle

In 1950 the Communist Party of South Africa (CPSA) became the first organizational victim of the Nationalist regime. By the late 1950s it must have appeared inevitable to all concerned that the days of the Congress Alliance's legal operation were numbered. The government used Sharpeville as a pretext for the battery of legislation that would ban the African National Congress (ANC) and make it all but impossible for the other national liberation movement constituents to operate. Denied opportunities for continued agitation in the public sphere, the Alliance affiliates and other organizations working to defeat the Nationalists and end apartheid turned to the contemplation of possibilities for action against the state, including armed struggle.

Members of the ANC, in concert with other individuals in the Congress Alliance affiliates, decided to form an armed movement, which, it was hoped, would prove the leading edge of an insurrectionary force that might ultimately challenge the government for power. It would not have been surprising if even the most dedicated members of the Congress of Democrats, faced with the disintegration of the legal national liberation movement, would have opted out of further activity, choosing either to leave the country or return to the anonymity of the suburbs, there to monitor events before deciding how to proceed. Rather, a number of Congress Of Democrats (and in many cases also SACP) members played integral roles in the formation of Umkhonto we Sizwe (MK), the ANC's new armed wing. These whites' academic and technical skills, their affluence, and the freedom, mobility, and access granted them by virtue of their whiteness, proved as valuable in the short-lived sabotage era as they had throughout the 1950s. Whites who had participated in the Congress Alliance were not alone in choosing to participate in these extra-legal and dangerous activities. Another organization formed specifically to engage in acts of sabotage, the National Committee for Liberation (NCL), a non-racial organization, harbored a number of former Liberal Party members among its underground membership.

The decision to move to violent forms of struggle was transcendent. It greatly increased the personal stake for those choosing to remain involved. For that reason, the fact that whites did voluntarily choose to take such risks and share equally in this new phase of the national liberation movement had a tremendous impact on the thinking of the Congress' African leadership, and on the shape of the ANC-led national liberation movement in exile. Members of MK, black and white, were risking their lives, and placing trust for those lives in each other's hands. These experiences, whether materially successful or not, left little doubt that whites, at least a handful of them, could be fully vested as comrades. This acceptance was a victory for those who believed that non-racialism had to be a significant, organic, and expanding component of the liberation process. But this victory was born of real blood and sweat. The Freedom Charter could evoke a vision of a non-racial democratic society, but if such a society was going to be built, it would be realized through many actions that would not necessarily describe a clean, unilateral line of progression. The participation of whites in the national liberation movement through COD was a step toward that vision. So too was white participation in the period of armed struggle and its aftermath. However, there were absolutely no guarantees, either of victory against the entrenched system or in realizing a non-racial society. One could state only, as Denis Goldberg described the meaning of his participation, that those whites who had chosen to stand up against apartheid had put a bit of coin down "toward paying the premium" that would have to be paid were a common society to be realized.

Sharpeville and the State of Emergency

By the late 1950s there was little evidence that South Africa was on the point of sweeping socio-political change. The Congresses were in a defensive posture, and their attempt to link with liberal allies did not necessarily overlay a revolutionary plan hatching outside the public eye. Although there were signs in 1959 and 1960 that the people, particularly in the rural areas, were becoming increasingly frustrated and unhappy with the ANC's policy of non-violence, the leadership of the ANC and the other Congresses had not come to terms with the effects that would flow from a turn to the use of violence. Did the initiation of sabotage rule out the further use of non-violence? What did a turn to violent activity mean in a state that enjoyed its own armaments industry and the strongest, most well-equipped military on the continent? There may have been individuals within the African National Congress by 1959 who were decisively in favor of a turn to violent struggle, but the organization as a whole exhibited a marked indecisiveness in its strategy, even after having turned to the armed option.

In 1959 the ANC decided to attempt a revitalization through a nation-wide anti-pass campaign. The campaign was to take place in the first half of 1960. The exact sequence of events remains controversial, but it is generally accepted that the Pan-Africanist Congress (PAC) had also been planning an anti-pass campaign and pushed up the date for the initiation of its campaign in order to get a jump on the ANC. The PAC campaign was to begin on March 21, when Africans would go to their local police stations in the townships without their passes and surrender themselves for arrest. The leaders of the PAC hoped that the expected influx of Africans would swamp the jails and make further pass arrests untenable. Africans began to gather around the police station in the Sharpeville township in the early afternoon. The crowd grew but was described by observers as peaceful. The afternoon's relative calm was broken by the report of a police revolver, which led to a full volley by the assembled police. Sixty-nine Africans were shot to death, many shot in the back, and 170 were injured. The Sharpeville massacre touched off a nation-wide barrage of protests, led to a two-week suspension of the pass laws, and ended in the imposition of a State of Emergency and the banning of the ANC and PAC.

In a memo dated the 22nd of March, 1960, and written prior to the Sharpeville shooting, the Congress of Democrats sent out a call to its allies to provide COD with a message of support for its upcoming National Conference:

> This conference is being held in Africa's year of destiny. The impact of great and rapid changes which are taking place throughout our continent is deeply affecting our political struggles at home.
>
> The Nationalist government, apparently so inviolable only a year ago, is beginning to show signs of serious dissension within its ranks. Apart from the significant changes taking place on the parliamentary scene, the discontent and anger of the great masses of oppressed Non-Whites is rapidly approaching a boiling point, as demonstrated by the recent flare-ups in Vereeniging, Cape Town, Durban, Paarl, Windhoek and the Transkei.
>
> In this fluid and tense atmosphere our organisation is holding its annual conference to discuss how best we can draw together all democrats into a unified force which can rapidly bring about our common aim -- the fall of the Nationalist government, and the utter defeat of all racialist ideologies.[1]

The year 1960 was to be a year of destiny for the national liberation movement in South Africa, but as a result of state, rather than popular, initiative. The COD memo found hope in an apparently fluid political landscape, both at the executive level and in Parliament, and among the urban and rural masses. The events following the shootings at Sharpeville, including the government's suspension of the pass laws, talk of change among government officials, and the appearance of mass protest and agitation among blacks, were interpreted by COD members as an outcome of this fluidity. The massacre, the mass protests immediately following upon it, and the imposition of the Emergency made organizing conditions all the more difficult. Many activists were detained, and others fled the country or were sent out. A number of COD members left for Swaziland. However, the five months of the Emergency were also one of the most intense periods of activity in Congress history for those members of the Alliance who were not imprisoned or did not leave the country:

> First of all we had to get together the people who were still around. We had to make contact with them and it meant that there were far fewer people around. When leaflets were produced, the people who were left had to do all that sort of work, they had to produce leaflets, they had to run them off and they had to distribute them. And so somebody had to do that work, somebody had to keep in touch with the various groups that were still around, which meant attending meetings. Those who were not in groups, those who were around singly, had to be contacted singly.[2]

Sharpeville proved to be a kind of wake-up call for the Congresses. At a time when the Congress Alliance had lost its lead organization, the ANC, to banning, low-level Congress members assumed new responsibilities. Catalyzed by Sharpeville, Congress sympathizers and members of the Congresses who had lapsed in their Congress work returned to the political scene. Michael Picardie, writer, theatrical producer, and some-time member of COD and the Liberal Party was jolted out of his political dormancy by Sharpeville. Picardie joined ANC couriers driving into Orlando, Alexandra, Moroka, Jabavu, and other Johannesburg townships distributing pamphlets:

> I was out one night with a man called Tennyson Makiwane. Like a fool I stopped near Alberton Police Station. The Police came out and they started to search the car, and they found thousands of pamphlets in the back of the car with "Chief Lutuli Burns His Pass," so we were arrested, the Special Branch was called; there was a scene in the cop shop where Tennyson asked for some water, and he was given a cup by a white policeman, and he drank from the cup and gave the cup back, and a policeman saw that he'd drunk from a white man's cup and smashed the cup on the floor.[3]

Surprisingly, Picardie and Makiwane were questioned, searched, and allowed to leave without the pamphlets. Only ten days later did the police come looking for Picardie. COD's first organized activity following the shootings was the release of a press statement that laid blame for the events of the two previous days on the police, the Nationalist government, and the white population generally. The statement rather threateningly suggested that it was no longer the white man's prerogative to determine South Africa's future.[4] COD's first discussion of the impact of Sharpeville and the Emergency took the form of a printed "discussion statement" in May 1960. The COD author described Sharpeville as the necessary outcome of African frustration and poverty. That the police had triggered the initial episode was not mentioned. The nature of the apartheid structure was invoked to explain "the basis of the crisis," and there was a sense of hope and forward movement in the fact that:

> An extraordinary number of prominent and representative people have recently expressed grave misgivings about the South African way of life, in some cases hesitantly and with many qualifications. Conservative businessmen, mining magnates, and ordinary people are questioning not just the way the Nats are handling our state affairs, but more fundamentally things like the pass laws, franchise rights, and the whole problem of the Africans' place in our society....all these expressions for a new way forward are to be welcomed.[5]

COD members sensed that a pivotal moment had arrived when white public opinion, driven by its first taste of the potential might of organized African anger and the censure of its international allies, was liable to make admissions of error and concessions in the treatment of the majority population. In its first post-Sharpeville leaflet, "Time to Think," COD set out a series of demands. The leaflet called for a government of concession and conciliation, the release of all detainees, the lifting of the Emergency, an increase in wages, and the lifting of the pass laws.[6]

With the end of the State of Emergency, the presidents of COD, the South African Congress of Trade Unions (SACTU), and the South African Indian Congress (SACTU) issued a press statement calling on the government to unban the ANC, repeal the Group Areas Act, end the carrying of passes, and institute a national minimum wage.[7] The government acceded to none of these requests. An objective examination of the State of Emergency's aftermath might have led Congress leaders to conclude that, organizationally, the national liberation movement had suffered a major defeat and was in crisis. However, interpretations following the Sharpeville period and the conclusion of the five-month State of Emergency were mixed. The ANC had been banned, new security legislation had been introduced, and many activists were either in prison or had fled the country. Still, the people had shown a willingness to rise up in their anger and frustration. South Africa had been roundly condemned in the international community. Most important, the economy had suffered an apparent body blow as worried international investors scurried to leave. In the immediate aftermath of the massacre, members of the business community, the United Party and liberals demanded that the government make concessions to the black majority as a step toward stabilizing the country.[8] It was possible to see the historical moment as one of opportunity, during which the state's weaknesses could be exposed and exacerbated. The people appeared to be at a point where they were willing to meet violence with violence. Yet the government refused to make any concessions beyond the temporary suspension of pass arrests and used whatever might was necessary to bring the situation under control. Did the Congress Alliance have, in its arsenal, an armed option to exercise? Could it afford not to in light of the events of 1960?

Whites and the Emergence of the Armed Option

Prior to initiating acts of sabotage, the ANC made a further attempt to win support for a national convention. South Africa was to be proclaimed a Republic on May 31, 1961, following a whites-only referendum. In late March, a broad array of representatives of the African people met in Pietermaritzburg for an All-In Conference called by the ANC. The delegates to the Conference demanded:

> That a NATIONAL CONVENTION of elected representatives of all adult men and women on an equal basis irrespective of race, colour, creed or other limitation be called by the Union Government not later than May 31, 1961; that the Convention shall have sovereign powers to determine, in any way the majority of the representatives decide, a new non-racial democratic Constitution for South Africa.[9]

Should the government ignore its demand, the delegates resolved to demonstrate against the Republic on the eve of its creation, to refuse cooperation with it in any manner, and to ask other nations to impose sanctions against it.[10] The Conference elected a National Action Council (NAC) to coordinate these protests, and invited participation by the Indian and Coloured communities and "democratic" whites in such protests if they proved necessary. The Congress of Democrats issued one of their final leaflets, "The Saracen Republic," calling on whites to support the demand for a National Convention as the only alternative to "bitter strife, suffering, bitterness and bloodshed."[11]

South Africa did become a Republic on May 31, 1961, and if COD's rhetoric was taken on its merits, the government had rejected the last peaceful option in so doing. The possibility of revolutionary violence was being contemplated at the time in a number of quarters. At its 1960 conference the South African Communist Party, which had reconstituted itself underground in 1953, instructed its Central Committee to devise a Plan of Action that would involve the use of economic sabotage. In 1961 a group of primarily white disaffected Liberal Party members and others formed the National Committee for Liberation with the purpose of engaging in sabotage activities. The Pan-Africanist Congress had also suggested that it might resort to violent methods of struggle, though at the time it was not making preparations for this. When did the ANC decide to form an organization to engage in sabotage, and how did the decision come about? In his speech from the dock prior to sentencing in the Rivonia Trial, Nelson Mandela stated that the decision to form Umkhonto we Sizwe was taken in the May-June 1961 period. MK was not formed as a wing of the ANC. It could not have been, given the need for secrecy around its formation and the impossibility of taking the question either to the ANC's leadership or to its followers. MK brought together members of the Congress Alliance who believed that it was now necessary to engage in violent activities if the movement for liberation was to be moved forward.

The core organizers of MK were strongly influenced by de-colonization movements in other parts of the world. Members of the South African Communist Party had embraced the notion that the majority of South Africans were experiencing "colonialism of a special type." Many of the leading theoreticians within the liberation movement thus chose to view the South African situation as inherently similar to anti-colonial struggles in other parts of the world. MK leaders, in formulating a plan of action, looked to and found lessons in the revolutionary theory and practise of others:

> I think the seminal work was Ché Guevara's *Guerrilla War*. And many of the errors can be attributed to that fact. Debray came later, but Ché certainly raised the notion. I remember the discussions at the time. The point made was, is it wrong to wait for the objective conditions to develop where violence becomes naturally part of the struggle, or is the situation such that the subjective element can create the objective, and that's Ché's argument. And we became convinced that the subjective could actually play an important role. And then the notion of an armed force arose. And then, what form should this violence take? And then people thought, well, in the first stage certainly we need to do cold acts of sabotage, and then hot acts, and then create a guerrilla army.[12]

A plan of action entitled Operation Mayibuye was formulated. The plan envisioned a series of sabotage acts as a means of priming the masses for armed insurrection. A core of 120 externally trained MK soldiers would return to South Africa and recruit cells of guerrilla warriors. MK's focus was to be on rural organization, and the Chinese, Cuban, and Algerian examples were considered illustrative of then-recent rural insurgency strategies.[13] How did the theoretical grounding translate into action? Denis Goldberg was an early member of MK. Goldberg was an engineer with a talent for technical matters. Goldberg taught first aid, electrical circuitry, telephone tapping, duplicating, and other such skills to MK recruits. Goldberg was also involved in fitness work with recruits, including twenty-five-mile fitness walks in the mountains behind Cape Town. There were also training camps organized for MK recruits:

> We organized a training camp at a place called Mamre, near Malmesbury [Cape Town region]. It was raided by the police. See, we made a mistake. Two mistakes. I bought the bread for the ten-day camp. It was put into plastic bags while it was still warm. It went moldy. So we had to buy more bread. And we went to a local village shop. And the shopkeeper wanted to know why we wanted so much bread. I didn't actually go. Somebody else went. "Where are you camping?" [he asked]. And in no time officials were there.[14]

Albie Sachs, son of E.S. "Solly" Sachs and a lawyer in Cape Town was at the Mamre camp that day giving a "class" in South African history. Recounting the day, Sachs remembered that they were in a very hot tent and some of the assembled dozed off, only to be awakened with a tree branch wielded by Looksmart Solwandle.[15]

These attempts at training -- physical conditioning and political education of a core of sabotage activists -- were amateurish in the extreme. Those involved had no means of getting to safe training areas and were constantly under threat of discovery. Nor were there many opportunities to practise actual demolition, which made every act of sabotage a trying experience. Joe Slovo, who was MK Chief-of-Staff, has been the most publicly prominent white member of MK. Slovo was born in Lithuania and emigrated to South Africa at the age of nine. He joined the National Union of Distributive Workers as a shop steward when only sixteen. By the later 1940s Slovo was a member of the Communist Party's Johannesburg District Committee. A founder member of the Congress of Democrats, he participated in all the major Alliance campaigns of the 1950s and worked closely with Nelson Mandela in the formation of Umkhonto we Sizwe. Commenting on the early sabotage period, Slovo ventured that the turn to sabotage "found us ill-equipped at many levels. Among the lot of us we did not have a single pistol. No one we knew had ever engaged in urban sabotage with home-made explosives."[16] Jack Hodgson, known as the "desert rat" for his North African experience in World War II, was considered the military expert among MK cadres. Hodgson was a member of MK's Johannesburg Regional Command, and Jack and Rica Hodgson's apartment became MK's Transvaal munitions depot:

> Jack and Rica's flat became our Johannesburg bomb factory. Sacks of permanganate of potash were bought, and we spent days with mortars and pestles grinding this substance to a fine powder. After December 16 [initiation of the sabotage campaign] most of our houses were raided in search of clues. By a stroke of enormous luck the Hodgson flat was not among the targets. Had the police gone there, they would have found that permanganate of potash permeated walls, curtains, carpets, and every crevice.[17]

The earliest devices involved a mixture of permanganate of potash, aluminum powder, and acid. The first two ingredients became explosive when catalyzed with a drop of acid. For a timing device, a small bottle containing acid was covered at its opening with a thickness of cardboard. To initiate the device the bottle was up-ended, and when the acid ate through the cardboard and dripped into the powdered mixture it would explode. Slovo planned to use such a device to blow up the Johannesburg Drill Hall, site of the Treason Trial, but was surprised while placing the bomb. Thinking quickly, Slovo was able to cover himself, leave the area, and return in time to stop the reaction from taking place. There were instances in which mistakes could not be righted, and a number of MK members died, the first as a result of a premature detonation. There were attempts by MK members to procure more sophisticated weaponry for sabotage purposes. Ronnie Kasrils and Eleanor Anderson were members of MK in Natal in 1962 when Joe Modise arrived from Johannesburg to evaluate MK's progress in Natal. Modise chastised the Natal cadres for insufficient progress and pointed to the Transvaal MK's procurement of dynamite as an example to be duplicated. It proved particularly useful to have white colleagues to call upon in this instance, as Kasrils and Anderson, informed of a roadwork project, drove out to the site with a picnic lunch and "sipped their Cokes" while reconnoitering the scene. Anderson noted the padlock on the dynamite enclosure and suggested that if the type and number of the padlock could be obtained, an identical lock could be purchased and the key would open the enclosure. Kasrils wandered over to the enclosure and noted the number and make of the lock. Anderson found its twin in a Durban store, and a week later Kasrils, Eric Mtshali, Billy Nair, and Mannie Isaacs drove out to the site, opened the gate, cracked the dynamite box locks and, as the Durban papers trumpeted the next morning, stole "HALF-A-TON OF DYNAMITE."[18] Kasrils discovered a few days after the heist that dynamite safety regulations included strictures against making sparks around dynamite, driving over 15 miles an hour while carrying dynamite, or storing the substance in any but cool, dry, well-ventilated conditions. The MK cadres had broken each of these strictures and in exact order.

Members of MK did enjoy some successes during the sabotage period, including destruction of a number of power pylons, telephone lines, government offices, and similar infrastructural targets. Arthur Goldreich, a white supporter of the national liberation movement who had studied guerrilla tactics while fighting the British in Israel, was involved in the drafting of Operation Mayibuye, which was offered as a blueprint for MK's movement from sabotage to guerrilla warfare. Goldreich also fronted for MK as a tenant when he bought a property known as Lilliesleaf Farm in the Johannesburg suburb of Rivonia. MK's High Command was set up on the Lilliesleaf property. On the 11th of July, 1963, the police raided the farm and found the bulk of MK's High Command there, poring over a draft of Operation Mayibuye. Walter Sisulu, Govan Mbeki, Raymond Mhlaba, Ahmed Kathrada, Lionel Bernstein, and Bob Hepple were arrested on the spot. Others joined to the trial included

Nelson Mandela, already in prison, Denis Goldberg, who was at Lilliesleaf at the time, Elias Motsoaledi, Andrew Mlangeni, Rusty Bernstein, Arthur Goldreich, and Harold Wolpe. Wolpe's brother-in-law, James Kantor had also been present, but was not politically involved and was discharged before the end of this, the Rivonia trial.[19] In their surprise raid the police found all the requisite documentation necessary for prosecution. The only question that remained at the time of the trial was whether the defendants would be executed. Goldreich and Wolpe, with Aesop Jassat and Moosa Moola, were able to bribe a young prison guard and escape South Africa. Bernstein was found not guilty but was shortly thereafter placed under house arrest, and he also managed to flee the country. The other Rivonia defendants were sentenced to life imprisonment. One of the leading saboteurs, Bruno Mtolo, known to the court as "Mr. X," turned state's evidence and proved a key witness for the prosecution.

Umkhonto we Sizwe had been brought to its knees in short order. Its members had underestimated the government's increasing intelligence sophistication. MK's activities may have created a degree of indignation among the white population and a heightened consciousness of African disaffection. However, the organization's activities did not succeed in creating a climate of anarchy and upheaval, because of its generally amateurish methods of operation, lack of technical sophistication, and dearth of materiel. In its initial manifesto, a flyer distributed at the time of December 1961 bombings, MK made clear its belief that armed struggle was to complement the non-violent mass politics of the national liberation movement. It was hoped that acts of sabotage would bring the government to rethink its position before bloody civil war was realized. MK's activities did not change government policies. Minister of Justice Vorster created a new arsenal of laws to use against dissenters, including the General Law Amendment Act of May 1963, which allowed for 90-day detention without trial. For the Congress Alliance, the SACP, and members of either who had entered MK, a discrete period of struggle had ended. What one Congress activist described in retrospect as the golden era of the 1950s had come to an end. The national liberation movement had come into its own, devised a charter, attempted to create a national structure of organization, and participated in a number of campaigns against the unfolding apartheid system. Non-violent resistance had been taken to its logical conclusion and had been banned. The armed option had been explored, taken up, and prosecuted within the knowledge and means of its progenitors. And the unity of black and white activists had been carried over from the Congress Alliance into the underground Communist Party and Umkhonto we Sizwe. Any legalist illusions as to the nature of the government had been stripped away. But the battle had been lost. The African National Congress and its allies would not re-emerge as a force to be dealt with inside South Africa for nearly fifteen years.

In the midst of the early 1960s' turbulence, members of the Liberal Party and other supporters of liberation were faced with much the same choices as the Congresses. Government repression and state law were willingly used against liberals, and it became increasingly difficult for those who had formerly considered themselves "safe" from government attack to operate as before. Thus there is a certain historical irony in the fact that the last acts of sabotage committed within South Africa in the 1960s period were carried out by an organization whose membership base included a number of Liberal Party members.

The National Committee for Liberation

The National Committee for Liberation (NCL), later known as the African Resistance Movement (ARM), engaged in acts of sabotage from 1962 through 1964. Monty and Myrtle Berman were among NCL's founder members and were involved in a number of acts of sabotage prior to the official foundation of the organization:

> He [Monty Berman] was actually involved in the first attack on pylons. And I mean, what we did at that stage, they went out with a hacksaw, and they went to a place in the Magaliesburg where they'd found a pylon up on a hill, and they spent several weekends returning to this place with a hacksaw and sawing through this thing which never fell.[20]

It is generally agreed that the NCL was a loose affiliation of a number of organizations, including the Socialist League, the African Freedom Movement,[21] and dissident Liberal Party members.[22] There were also members who had had little experience politically prior to joining the NCL. Some of these were at university at the time. The NCL brought together Trotskyites and anti-communists. Among the most prominent members of the NCL were Monty and Myrtle Berman, Randolph Vigne, Baruch Hirson, Robert Watson (who had some knowledge of explosives), and Adrian Leftwich, a student at the University of Cape Town. The NCL was a non-racial organization. Although the preponderance of its members were white, there were African and Coloured members also. However, with the exception of Eddie Daniels, it appears that the decision-making core was white. The NCL was divided into regional commands, one each in Johannesburg, Durban, and Cape Town. There were rare meetings that brought together representatives of each of the regions, but generally members in one region did not know the members outside that region. Pseudonyms were used to promote security. Hirson claims that the NCL was led from Johannesburg. However, as Hirson acknowledges, the Cape Town region's activities received more attention than those of the other regions.

The NCL, like MK, had a conscious policy of respect for human life in its sabotage planning. Therefore, NCL focused on infrastructural targets whose destruction would not endanger people. These targets included power pylons, a number of which were successfully downed, a radio transmitting structure, and the Cape southern suburbs rail signal system. While the NCL did have in its service a few individuals with some experience in demolition, the members charged with carrying out acts of sabotage made mistakes much like those of their MK counterparts, including the storage of dynamite in hot and humid conditions. Lewin noticed once, while transferring dynamite into plastic bags, that it appeared to be "wet," a signal that dynamite has become self-explosive. Hugh Lewin's statement below is indicative of the lack of preparation and training that most NCL members brought to sabotage work:

> I never thought that I would be in a position to do this. I never thought that I would go to jail. I was white, living in the nice, easy white society, and even when I started to do most un-white sorts of things, like blowing up pylons, it was possible to slip back into the white suburban ways, which provided a useful screen against detection -- and a screen against unwholesome reality.[23]

Lewin had worked with the Liberal Party prior to being approached by a member of the NCL who asked him to join. It may not have been clear to Lewin at that initial moment that he was joining an organization rather than simply a loose band of activists engaged in "protest sabotage."[24] During his eighteen months in the NCL Lewin engaged in three acts of sabotage with a team of three other members, Fred Prager, Baruch Hirson, and Raymond Eisenstein.[25]

The last pulse of NCL activity came at the time of the verdict against the Rivonia defendants. Some time between January and April 1964 a national meeting of NCL representatives was held in Johannesburg.[26] At that meeting the status of the NCL's sabotage operations was discussed and assessed. The suggestion was made that the NCL should form a political wing purely for propaganda purposes. Among those in attendance were a number of NCL members who were either strictly against continued sabotage activity or were wavering on the question of whether NCL should continue. Adrian Leftwich, President of the National Union of South African Students (NUSAS) and a leader of the NCL in Cape Town, sought to have Eddie Daniels removed from the organization, fearing that Daniels' strong anti-sabotage stand might lead others in the group to rethink the sabotage question. Leftwich was counseled against speaking for Daniels' removal, and it was decided that the group would continue with acts of sabotage but would focus on targets with impact but with greater security for the participants, being more remote than some previously chosen targets.[27] Shortly after the Rivonia verdicts, and in preparation for a post-Rivonia directed series of sabotage acts, an organizational statement was drafted in Johannesburg to be released to press organizations at the time of the bombings. This was in mid-1964, and at this time the organization presented itself as the African Resistance Movement (ARM). The statement noted ARM's dedication to the overthrow of the apartheid system and the creation of a democratic society "in terms of the basic principles of socialism."[28] The statement's drafters saluted the Rivonia defendants and stated ARM's desire to see the unification of the anti-apartheid organizations currently operating within the country. The language of the statement was militant:

> ARM does not only talk. ARM acts. ARM has acted. ARM has declared and will declare itself through action. This is the only language our rulers understand. And ARM, with other freedom forces, will harry and resist the oppressors until they are brought to their knees.
>
> White South Africa has often been given the opportunity to align itself with progress. It has constantly refused to do so. It has sought only to build for itself, on the backs of the people, a comfortable bastion of profit, power and privilege.
>
> ARM declares its fight not against the whites as such but against the system they jealously defend. ARM will avoid taking life for as long as possible. ARM would prefer to avoid all bloodshed and terrorism. But let it be known that if we are forced to respond to personal violence -- and we cannot forget decades of violence, torture, starvation and brutality against us -- we shall be prepared.[29]

ARM's bark was certainly meant to be worse than its bite. However, the final act of sabotage, which was considered to be an ARM act yet came two weeks after the organization had disbanded, led to the loss of a life. John Harris planted a bomb in the corridor of the white section of the Johannesburg rail station, which left one woman dead and many others seriously injured. Hugh Lewin remembers the day clearly. Lewin was taken from his cell, driven to the scene of the explosion, brought back to Special Branch headquarters, and severely beaten. Harris's act had not been planned or sanctioned by the ARM leadership, but because it was considered an ARM act, each member of the group was considered implicated in Harris's action. Harris was convicted of murder and hanged five months later.

On the 4th of July, 1964, the police began a series of raids that netted a number of ARM members. Hugh Lewin and Adrian Leftwich were among those arrested. In Leftwich's apartment the police found copies of many ARM documents. Lewin noted that Leftwich probably knew more names of members than anyone else in the organization. And when he was arrested and interrogated, he gave the police the names of other members. Among the documents the police found in Leftwich's apartment was a "Proposal for Addis Ababa," which criticized the organizations engaging in sabotage for their unwillingness to work together. The document called for the creation of a "joint command" to be set up among those organizations.[30] A number of other documents commented on the current South African political situation and the nature of revolutionary possibilities: "The thesis thus far is this: as yet in South Africa there are not even the most essential and limited conditions for effective guerrilla warfare and the present sabotage phase has no real effect militarily, and stands a very good chance of being smashed. All it can hope to do is affect the climate."[31] The ARM member or members responsible for the creation of these documents shared a number of tendencies with their counterparts in MK. Both were aware of and appeared to incorporate some aspects of Ché Guevara's thesis on the creation of revolutionary conditions (Baruch Hirson wrote a précis of Guevara's work for ARM).[32] Members of both organizations also looked to other revolutions or insurrectionary situations for parallels and possibilities; ARM documents looked to Algerian and Chinese revolutionary examples.

There were a number of trials of ARM members in the latter half of 1964. Lewin and Eisenstein received seven-year sentences. Hirson received a nine-year sentence, and Praeger was acquitted. Albie Sachs defended two ARM members in a trial with five defendants and six listed co-conspirators. Sachs represented Alan Keith Brooks, then a junior lecturer at Cape Town. Speaking in hopes of mitigation of sentence, Brooks told the court that he had joined the NCL out of a feeling of frustration. He pointed to the promotion of apartheid within the universities, the Sharpeville debacle, and the country's departure from the Commonwealth, with the subsequent institution of a Republic in 1961. Brooks considered that "[A]ll these events indicated to me that normal lawful opposition was futile and it might be necessary to resort to extra-legal means of pressure."[33]

There were other members of NCL, including Tony Trew and Stephanie Kemp, who had been only peripherally involved in politics prior to joining the NCL yet who now faced prison sentences. Both MK and the ARM were brought to heel by the government in 1964. Both organizations acted, admittedly, out of a sense of frustration, and with a desire to shake up the government and the white population while neither repudiating non-violent protest nor hurting human beings. Both organizations enjoyed some small measure of success and both learned hard lessons as a result of the amateurish nature of their work. However, MK's grounding in a larger structure, the Congress Alliance, imbued it with a history, a purpose, and a link to survival that the NCL never enjoyed. Of the two organizations, MK appeared capable of learning from its mistakes. There was an expectation that there would be a next time. ARM had been a very loosely knit group composed of a number of basically autonomous organizations that shared a sense of frustration and a belief that non-violence was no longer a realistic or sufficient answer to government initiatives. ARM's members shared no common history, were from diverse ideological backgrounds, and for the most part were not even aware of each other's membership in the organization. Repudiated by the Liberal Party, ARM's brief history ended with the Fourth of July raids.

Albie Sachs, lawyer and Congress supporter, could well have been speaking for many white activists when he wrote these lines, at the time of the ARM trials, musing on the probability of his own arrest:

> What can they say of me? You urged whites not to close the door on non-white demands -- and plead guilty. You tried to explain to whites why non-whites were demonstrating -- guilty. You said that arming the whites to fight the non-whites was pushing the country along the road to disaster -- guilty. You said that negotiation was better than force, that negotiation was possible, that a national convention representative of all races should gather soon to hammer out a new constitution for South Africa -- guilty. You said and you felt that there is a secure place for the white man in this country, but only when all its inhabitants have full citizenship rights -- guilty. You believe in the essential equality of man and in human brotherhood -- guilty. Yes I am guilty, and proud of it, of propagating all these ideas -- but so what? Can that possibly make me a criminal, even in this country where the right to protest and dissent is so severely restricted?[34]

Beyond the divisions, ideological and organizational, that separated COD members from Liberal Party members, that kept socialists apart from communists and that kept some communists out of the Congress Alliance,[35] there was a common denominator among the "white democrats" that transcended other real differences. This commonality could not have been phrased more eloquently than it was by Albie Sachs. It was a commonality of identification based upon a belief in the "essential equality of man" and the abhorrence of racial division as manifested in apartheid. It was the desire to break down those barriers in order to work with, and to be seen by the oppressed as committed to, a just and democratic alternative. The barriers that white activists broke down were as much barriers to their own freedom, and to their growth as human beings, as they were barriers to their black counterparts' realization of their rights. Ultimately, South Africa's white radicals were liberating themselves.

Notes

1. Congress of Democrats, Call for messages to COD Annual National Conference, 22 March 1960, signed Ben Turok, M.P.C., SAIRR, Political Documents, Microfilm 860, Institute of Commonwealth Studies.

2. Interview, Ester and Hymie Barsel, conducted by Julie Frederikse, held at the South African History Archive, Johannesburg.

3. Interview, Michael Picardie, London, 24 January 1987.

4. Congress of Democrats, "Press Statement," 23 March 1960, SAIRR, Political

Documents, Microfilm 860, Institute of Commonwealth Studies.

5. Congress of Democrats, "Discussion Statement," May 1960, SAIRR, Political Documents, Microfilm 860, Institute of Commonwealth Studies.

6. Congress of Democrats, "Draft Statement for the Acting National Committee of COD," May 1960, SAIRR Political Documents, Microfilm 860, Institute of Commonwealth Studies.

7. Press statement, 6 September 1960, signed by Dr. G.M. Naicker, President, South African Indian Congress, Mr. Leon Levy, President, South African Congress of Trade Unions, Mr. P.A.B. Beyleveld, President, South African Congress of Democrats, SAIRR Political Documents, Microfilm 860, Institute of Commonwealth Studies.

8. Karis and Gerhart, Vol. 3, *Challenge and Violence*, 336.

9. "The Full Text of the Resolution of the Pietermaritzburg All-In African Conference, held on March 26 & 27, 1961," *Fighting Talk*, Vol. 5, No. 3, April 1961.

10. Members of the Pan-Africanist Congress who were originally going to participate in the actual Conference declined, stating that they were prepared to work for African unity but were not prepared to work toward a multi-racial National Convention providing representation to all parties including whites. Brian Bunting, "Towards a Climax," *Africa South*, July-September 1961, Vol. 5, No. 4.

11. "The Saracen Republic," (leaflet), Congress of Democrats, 1961, Colonial World Pamphlet Collection, Institute of Commonwealth Studies.

12. Interview, Ben Turok, London, 27 January 1987.

13. Colin Bundy, " 'Around Which Corner': Revolutionary theory and contemporary South Africa," *Transformation* 8 (1989), 5. Bundy notes that Mayibuye was divisive within MK. A number of leading members thought the plan was ridiculous.

14. Interview, Denis Goldberg, London, 3 February 1987.

15. "The Least Dramatic Contribution," Albie Sachs in *Dawn: Journal of Umkhonto we Sizwe*, 1986. Looksmart Solwandle was detained a year later and was found dead in his cell, allegedly of suicide by hanging.

16. "The Longest Three Minutes of My Life," Joe Slovo in *Dawn: Journal of Umkhonto we Sizwe*, 1986.

17. Ibid.

18. "Dynamite Thieves," Ronnie Kasrils in *Dawn: Journal of Umkhonto we Sizwe*, 1986.

19. See Kantor's account of the period, *A Healthy Grave* (London: Hamish Hamilton, 1967).

20. Interview, Hugh Lewin, Harare, 10 August 1988.

21. The nature of the African Freedom Movement is unclear. Baruch Hirson, who came to the NCL from the Socialist League, claims that AFM members were mostly members of the Congress Youth League, and that the organization had been formed during or after the State of Emergency. Hirson had known most of them prior to that time during his short tenure in the Congress of Democrats. Hirson notes that the members of the AFM he met with stated that they had 120 members. Apparently there were also members of MK who engaged in acts of sabotage as members of, or with explosives provided by, NCL. Interview, Baruch Hirson.

22. John Marcum and Allard K. Lowenstein described the NCL as including "socialists, disaffected members of the Liberal Party and of the Congress of Democrats, some Trotskyites, and some persons not known to have been previously involved in

politics. A large proportion of the group was white, including a number of demolition experts....it was hostile to cooperation with Communists." Marcum and Lowenstein in John Davis and James Baker, eds., *Southern Africa in Transition* (London: Pall Mall, 1966).

23. Hugh Lewin, *Bandiet: Seven Years in a South African Prison* (London: Heinemann, 1974), 15.

24. Interview, Hugh Lewin, Harare, 10 August 1988.

25. Lewin, *Bandiet*, 13.

26. This account is based upon the corroborating remembrances of Hugh Lewin and Baruch Hirson, both of whom were at the Johannesburg meeting. Lewin remembers the meeting as being around Easter, in March or April. Hirson remembers it as a January meeting.

27. Interview, Hugh Lewin, Harare, 10 August 1988.

28. African Resistance Movement, statement, n.d., ARM File, Institute of Commonwealth Studies.

29. Ibid.

30. Court Document C4, "Proposal for Addis Ababa," ARM document, n.a., 6 July 1964, Albie Sachs Papers, Institute of Commonwealth Studies.

31. Court Document C7, untitled, n.a., n.d., Albie Sachs Papers, Institute of Commonwealth Studies.

32. Interviews, Baruch Hirson, London, 8 January 1987, 20 May 1988.

33. Alan Keith Brooks, "Statement in Mitigation," South African court transcript, n.d. Albie Sachs Papers, Institute of Commonwealth Studies.

34. Albie Sachs, handwritten document, Albie Sachs Papers, Institute of Commonwealth Studies.

35. During the "One Congress" question debate in COD, the non-participation of some white communists in the Alliance because of its federal structure was intimated. The author has heard this from interviewees as well.

11

A 1990s' Epilogue: Non-Racialism in the Final Analysis?

To the casual observer comparing the multi-racial protest politics of the 1950s and the renaissance of mass-based, multi- and non-racial organization in the 1980s, there may have been a number of apparently striking similarities. The resurgence of mass protest, spanning the spectrum of local and national issues and involving black and white in common activities, was immediately remarkable. The Freedom Charter once again was publicized and rose to prominence as one of the central statements of the national liberation movement inside South Africa. It might have appeared easy, in 1983 or 1985 or 1989, to conclude that there was some unbroken historical tradition connecting the two eras -- that recent victories reflected the struggles of past decades described here in Chapters 1 - 10. This chapter will examine the basis -- or lack of it -- for any such conclusions through the early 1990s.

A number of organizations promoted non-racialism in the 1980s. The United Democratic Front (UDF) was established in 1983 as a non-racial umbrella structure uniting hundreds of diverse people's organizations; bulwarks of the 1950s period, including Helen Joseph, were present at the Front's launch. As anti-apartheid activists became bolder in the 1980s, the presence of the banned ANC, its colors, literature, and evidence of its *sub rosa* activity, became increasingly visible. Ethnically and racially based organizations like the white Johannesburg Democratic Action Committee (JODAC) and the reborn Transvaal Indian Congress (TIC), organizations similar to affiliates of the Alliance in the 1950s, were founded or re-formed to bring the message of the national liberation movement -- and in many minds the African National Congress specifically -- to target audiences. The rhetoric of the 1980s was replete with references to the creation of a non-racial, democratic South Africa, echoing the language of the Congress of the People and the Treason Trial. One might have been moved, in 1983 or 1985 or 1989, to draw strong parallels between the 1950s and the 1980s, even to claim a victory for non-racialism as evidenced by the Freedom Charter's durability, the rhetoric of the UDF and the Congress of South African Trade Unions, and the will to unite and create non-racial organizations in the face of government repression far greater than that experienced thirty years earlier.

However, any conclusion that non-racialism was, by the end of the 1980s, victorious, in the sense that it was an unquestioned tenet of the liberation movement and a guaranteed fixture of a post-apartheid South African culture, must be held highly suspect. If non-racialism was victorious, the questions remained: Who was responsible for this victory, and for whom was it won? How did one measure this victory, and what would be its tangible benefit? A more studied examination contrasting the anti-apartheid activities and players of the 1950s and 1980s might suggest that differences between the two periods were as substantial as any similarities perceived, perhaps even greater. The activists of the 1950s were, in a sense, starting from scratch. They were reacting to, building theories based upon, and shaping tactics related to an apartheid structure that was then under construction, constant elaboration, and refinement, both technically and ideologically, throughout the 1950s. The activists of the 1980s, some of whom had also been the activists of the 1950s, had thirty years in which to weigh and analyze the terrain of struggle. Where the activists of the 1950s had begun from a base of legality, their 1980s counterparts worked under the opposite conditions: The legalization of the people's political organizations in 1990, rather than any measure of state repression, was cause for great surprise and, one might surmise, consternation among them, as they now had to reconstitute their activities and come to grips with the realities of transitional politics.

Even though the trappings of the UDF may have evoked faint recollections of the Congress Alliance, the differences were manifest. The Congress Alliance was, as its name suggested, a variably cohesive alliance of distinct organizations, African, Indian, Coloured, and white, that both promoted contact across lines of race and ethnicity and arguably heightened those distinctions by utilizing and magnifying them for purposes of organization. The Congress Alliance had been led by the African National Congress, a distinct African political organization. In the years between the banning of the ANC and its Congress Alliance partners in 1960 and the return of mass organization in the mid-1980s, the ANC had gone into exile and had taken decisions there that changed it greatly. By the time of the UDF's formation in 1983 the ANC was truly a non-racial organization. The ANC, at its 1969 conference in Morogoro, had opened membership to those of all races and ethnicities. At its Kabwe conference in 1985 the ANC voted members of other racial and ethnic groups, including whites, onto its National Executive for the first time in its history. Thus those members of the Alliance affiliates who had remained activists, many of whom had ultimately gone into exile themselves or were imprisoned, ceased being members of the South African Coloured Peoples Organization (SACPO) or the Congress of Democrats (COD) or the South African Indian Congress (SAIC) and became ANC members. The UDF, rather than serving as a shadow Congress Alliance, was a distinct consortium of organizations, many of which looked to the ANC as the acknowledged leader of the struggle for national liberation, but which maintained independent organizational lives, often with specific issue orientations.

The fact of the ANC's banning and exile between 1960 and 1990 is crucial in thinking about the role of white South Africans and the status of non-racialism in the period up to, and beyond, 1990. The ANC was not without influence inside the country during this period. Even after the highly repressive early 1960s, ANC cadres remained in the country. It appears that the ANC sought actively to rebuild its networks inside the country from the early mid-1970s. From the late 1970s ANC cadres who had gone into exile began infiltrating the country, and Umkhonto we Sizwe undertook an extended if sporadic campaign of sabotage activities. By the 1980s the ANC had created a sophisticated and extended external operation with a large propaganda arm. Still, the ANC was not allowed to forge a highly visible presence in South Africa from the early 1960s into the 1980s -- nor did it attempt to do so. The fact that the ANC was largely not present during the political developments of the later 1960s, 1970s, and early 1980s meant that a new generation of activists and organizers, workers and students had to feel their way without the direct guidance of the ANC. And they created organizations and structures separate from the ANC, if ultimately in synch ideologically and programatically with it.

This chapter will examine white participation in anti-apartheid and related activism from the early 1970s through the renaissance of mass protest in the 1980s to the early 1990s. As a new generation of African activists grappled with their history and their role in history-making from the late 1960s, a new generation of whites, most with no link to their 1950s forbears, also began to grapple with their role and responsibility in the maintenance or destruction of apartheid. A number of different, even contradictory, analyses arose from this reflection and activity. Whites, like all activists coming to grips with South African realities and future visions, sought to define their role as change agents, and to grapple with their relationship to non-racialism: as a vision, an ideology, a tactic and strategy, and perhaps at times and in certain places even as an imperfectly realized reality. Whites in the last decades of the twentieth century did face questions similar to those their counterparts had faced thirty years earlier: What place could whites find in a struggle for the liberation of a black majority? Who should whites organize, or who did whites have the right to attempt to organize, and was there a place for whites, as activists, in black areas? Was it right to maintain racially and ethnically distinct organizations while working for a non-racial society? The questions white activists working in the 1970s onward were asking may have been similar to those posed by and for their counterparts in the 1950s, but the conditions under which they were posed were quite different. Would the generations of the 1970s and 1980s find answers thirty years on?

White Activism in the 1970s

Many of the white South Africans who participated politically from the 1970s onward cut their political teeth as university students. The National Union of South African Students (NUSAS) was probably the most prominent forum for white student activists in the 1970s. In 1969 Steve Biko had headed the effort to break black students' relationship with NUSAS and their dependence on NUSAS as a mouthpiece for black students' interests. The result of Biko's efforts was the formation of the South African Students Organisation (SASO), which was sharply critical of NUSAS and of the role of white liberals and white liberalism generally in stunting black political development.[1] Nevertheless, whether members of NUSAS or not, there were young whites seeking the means to manifest their displeasure with apartheid. As one white South African who had been attending a university in the early 1970s put it, he had "felt that the majority of poor people needed to become involved in some sort of process of struggle, and that we [whites] needed to identify with them in that process [and] try and find a way of becoming part of that struggle."[2] With no visible structures of the national liberation movement in sight, and largely cut off from contact with black students, white students found no readily available political outlet for their energies and concerns.

That changed, to some extent, with the creation of the first Wages and Economics Commission at the University of Natal, Durban in 1971. Richard Turner, a young lecturer in Political Science at the University, and an advisor to NUSAS there, was largely responsible for this development. Turner had taken a B.A. at the University of Cape Town and then traveled to Paris, where he wrote a doctoral thesis on Sartre at the University of Paris. Turner wrote and lectured on South Africa while at Durban, and he offered his students a strongly materialist analysis of apartheid, elucidating the relationship of apartheid to capital accumulation.[3] Turner's analysis, describing race as an adjunct serving the interests of capital in perpetuating apartheid, was highly influential among a group of students at Durban who would form the first Wages Commission. The Wages Commissions researched workers' conditions and wages, published pamphlets and newspapers directed at workers, organized worker education programs and worker benefit societies, and served as launching pad organizations for a number of students who would become directly involved in union organizing.[4] Various of the organizations and advice bureaus formed by the white students of the Wages Commissions became precursor bodies to the Federation of South African Trade Unions (FOSATU), which was the first self-described non-racial union since the South African Congress of Trade Unions (SACTU), which affiliated to the Congress Alliance in the 1950s.[5] Alec Irwin, a lecturer at Durban involved in the Institute of Industrial Education there, ultimately left his university post to become general secretary of FOSATU. Irwin had been drawn to union work because he felt that "workers were central" and that change in South Africa would have to take place through economic restructuring and the end of underdevelopment.[6]

The rise of non-racial trade unions from the 1970s onward raises important questions about the developing meaning of non-racialism for different groups of people working in different environments and focused on different concerns. Irwin noted that the non-racialism of FOSATU from the time of its inception was "completely different" from the non-racialism of the UDF.[7] The non-racialism of FOSATU and that of the Congress of South African Trade Unions (COSATU) from the mid-1980s repudiated racial barriers of any kind within unions. The unions shared a common ground, that of the worker, which on one plane facilitated the transcendence of racial- or ethnic-based thinking and definition and on another plane allowed the participants to focus on, analyze, and understand those racial and ethnic divisions without treating them as primary determinants of the South African socio-economic formation.

The white intellectuals coming out of the Wages Commissions, the Institute for Industrial Education, and other worker-focused bodies generally brought highly theoretical backgrounds with them as they entered positions in unions whose membership was predominantly black. While they played contributory roles in building unions in the 1970s, those roles were problematic. They had to find, through trial and error, the parameters of viable white involvement. There was a fine line between using whites' advantaged positions in society (affluence and access) to union benefit and impeding black workers' rise to leadership by assuming that whites were somehow more capable or fit in leadership roles, from the shop floor upward, even if those whites lacked practical experience. This process was critical in helping unions define the relationship between worker struggles and political struggles, and the place of non-racialism in both.

A hallmark of the white union activists of the 1970s was their generalized hostility to the Congress tradition. Though often working with people who came out of the Congress and SACTU tradition, these whites were disdainful of it largely for the same reasons members of the Communist Party of South Africa (CPSA) had berated the ANC and held the politics of alliance in the later 1940s in contempt, even while participating in it. Change would come about in South Africa only if the society were overhauled at its roots, the point of production.

While the whites engaged in union work were the more prominent group of white activists in this period, there were other whites in the 1970s who were seeking to uncover the Congress tradition and find a place within it. A handful of young white South Africans sought in the 1970s to learn about the African National Congress, the Congress Alliance, and the SACP, and to make contact with the ANC.

Jeremy Cronin was one of those whites. His first political involvement took place at the University of Cape Town, where he joined the Students Radical Movement, choosing consciously not to get involved in NUSAS. Attempting to uncover the history of the national liberation movement in the 1970s was a difficult task; information was hard to come by. Cronin noted that:

> Occasionally someone would find a book in a secondhand book shop written by Govan Mbeki or Eddie Roux's *Time Longer Than Rope* -- it's now banned -- there were one or two copies around and people would copy them.... You knew they were worth their weight in gold.[8]

Cronin described his political conversion as "completely intellectual," since he had had no black friends. Grappling with his place in apartheid society, Cronin sought through his covert reading to find signs that there was a place for him among those working for change. The pamphlets and other readings he came across gave him hope, in that Congress materials generally made glancing mention of the fact that the struggle for a free South Africa was non-racial and there was a place for democratic whites there.

Finishing his studies in 1971, Cronin went abroad in an attempt to offer his services to the ANC. He was able to make contact with the ANC, and although he was welcomed he was also told to consider his motivations for involvement in liberation activities: He was not to seek involvement out of a sense of guilt or romanticism. He was instructed to return to South Africa, there to have no contact with his former radical friends or blacks, and to create a new "careerist" persona, which he did. Cronin and a number of other whites undertook propaganda work for the ANC in the mid-1970s, posting leaflets and setting off "bucket bombs," harmless devices that could be left in crowded urban areas to explode, releasing hundreds of leaflets.

Cronin was arrested in 1976 and the state attempted to use his trial to prove that white Communist Party members were behind the Soweto uprising, a claim with no anchor in reality. Cronin's trial was one of the first ANC trials in the period following the so-called "lull" between the early 1960s and the renaissance of activism and organization in the 1970s. Cronin noted that in later years black colleagues who had been involved in Black Consciousness organizations at the time told him that the trial, and the fact that a white was willing to go to prison for taking part in the national liberation movement, raised in their minds the possibility that it might be in the interest of Africans to forge alliances across racial boundaries -- "that it wasn't just a straight black-white struggle."[9]

Others followed Cronin in the 1970s; Barbara Hogan's case is probably the best known. Hogan began involvement with the ANC in 1977. She established a network to facilitate communication with the ANC outside the country, and between 1978 and 1981 she traveled to Botswana a number of times for meetings with the ANC-in-exile. Her work included the transmission of ANC documents to London. When Hogan was arrested in 1981 and tried, she stipulated to her membership in the ANC and was given a ten-year prison sentence. White South Africans played a part, if a small one, in the ANC's return to greater prominence and involvement within South Africa in the 1970s.

The number of whites participating in union work or in the underground ANC structures in the 1970s was minuscule. There were few venues for participation at this time. Students could find a role in NUSAS or some other campus-based organization. Others found a role in Beyers Naudé's Christian Institute, which the anti-apartheid clergyman founded after being expelled from the Dutch Reformed Church in 1963. In the early 1970s the Christian Institute funded publications that were disseminated among whites asking them to think about how they had come by their privileged positions in South African society. The Institute also worked with white groups to build links across racial and ethnic lines. This program was discontinued prior to the Institute's banning in 1977.

The 1970s ended on a revolutionary note. A new generation of black South Africans had risen in protest against the apartheid regime, had learned important organizational lessons, and had tasted their potential power. In doing so they had set in motion a process that would revive anti-apartheid activism within South Africa and lead to the formation of new organizations, drawing on the lessons of generations of experience, to take the challenge to the state to a higher pitch. The post-Soweto student organizing experience is illustrative. Two student organizations, the Congress of South African Students (COSAS) primarily organizing students at secondary schools, and the Azanian Students Organization (AZASO), organizing university students, were formed in 1979.[10] While both were initially oriented to Black Consciousness, each changed its ideological emphasis (COSAS adopted the Freedom Charter in 1980 as its program for a free society), shedding Black Consciousness in favor of an analysis that described the enemy, not as whites per se, but as exploitation under capitalism. This change in outlook can be explained in a number of ways. The increasing awareness of and discussion about the ANC, with its historical weight and increasingly active presence, and the prominence of the Freedom Charter as the ANC's statement of purpose must have been an important factor in moving youth away from Black Consciousness and toward acceptance of the doctrine of non-racialism. Some who would take leading roles from the late 1970s and early 1980s came into direct contact with ANC members in South African prisons or in exile, and that contact was often highly compelling. And the fact that a few whites, the Cronins and Baskins and Hogans, had taken risks in the name of national liberation and were paying a price for it, may also have been a factor in the discussions that took place on the road from the Black Consciousness of the 1970s to the inclusive non-racialism of the 1980s. Politically active blacks were also aware that the state had appeared to play a role supportive of Black Consciousness thinking prior to the Soweto uprising, which led to the banning of all Black Consciousness organizations in 1977. Black and white students undertook united activity in 1980 for the first time since black students had walked out of NUSAS. COSAS and NUSAS jointly orchestrated the 1980 Freedom Charter campaign, bringing the Charter to many communities that had not seen it since the 1950s. Black and white students collaborated in a number of actions from 1979 through 1981, including the boycott of Fattis and Monis food products and of the sweet maker, Rowntree. White students at the University of Cape Town worked with students at the University of the Western Cape, a radicalized black university, on the Fattis and Monis strike, which opened up channels of communication between white students and black communities. White students' participation in the Rowntree boycott was useful, given their access to printing facilities, but that participation remained problematic for a number of reasons. Community groups distrusted the presence of whites; the distrust led to the decreased participation of black community organizations. White students also had a tendency to take control of certain processes, such as pamphlet writing, which might have been better left to community members. One white participant noted that "as a result, white democrats in Cape Town to some extent controlled both the mental and manual side of a key aspect of the boycott."[11]

The years of the later 1970s and early 1980s were a time of building bridges and rebuilding trust among and between communities. While this study focuses on the relationships between black and white, the various forces ranged against apartheid in each community, be it African, Indian, white, or Coloured, had to overcome fears, animosities, and distrusts of each other that had built up over years of diminished contact and little collective action in any sphere. Whereas non-racialism as a doctrine had evolved almost organically out of the experiences of activists in the 1950s, the non-racialism emerging in the late 1970s appeared to be more an article of faith.

The acceptance of non-racialism was, in part, a legacy of South African political prisoners and exiles who had been silenced by the bannings of organizations and individuals. Also, that legacy had been all but effaced through the criminalization of literature relating the history of the ANC, the adoption of the Freedom Charter, and the imprisonment of the people's leaders. Blacks accepted non-racialism for what they perceived as its correctness in the larger firmament of correct principles of struggle (as expounded in the Freedom Charter, for instance); these principles were supported by activists who had lived it in the 1950s or who had learned it from those activists in prison or exile. In this sense, non-racialism came to have a life somewhat apart from the actions, or non-actions, of white people.

The Revival of Mass Political Organization

From the time of the Soweto uprising and related popular initiatives in 1976 and 1977, the state's actions suggested that it was cognizant of pending threats to white supremacy and planned to take the offensive against them. The state's strategy dictated that it use its powers to divide the African population. It did so in two ways: ethnically, through the consolidation of the bantustan system and the provision of "independence" therein while foreclosing on the South African citizenship of Africans; and through the creation of a more economically stratified African population. The state increased the rights of a settled urban, middle-class African buffer group while making it much more difficult for the majority of Africans to remain in urban areas.

These tactics of divide-and-rule were not confined to the African population. The government's initiative to create separate parliamentary chambers for Indians and Coloureds while greatly increasing and centralizing power in the hands of a state president was largely responsible for sparking the creation of the United Democratic Front. Rather than succeeding in coopting the majority of Indians and Coloureds with the promise of limited parliamentary representation, the campaign against the "Constitutional Proposals" led to the creation of the most important alliance across racial and ethnic lines since the Congress Alliance of the 1950s. The UDF's declaration stated that the Front committed itself "to uniting all our people wherever they may be in the cities and countrysides, the factories and mines, schools, colleges and universities, houses and sports fields, churches mosques and temples, to fight for our freedom."[12] True to its declaration, the UDF was composed of literally hundreds of sports, religious, workers, and student organizations from the whole of the country.

The UDF's formation was very important for whites seeking an outlet for their anti-apartheid energies. Organized as a non-racial front embracing the values of the Freedom Charter (although not adopting the Charter itself), the UDF appeared to open up a world of opportunities for contact, in common cause, among all South African communities. The UDF's formation also posed a challenge to whites who sought involvement. The alliance of organizations in the Front was nothing like the Congress Alliance, where racially and ethnically delineated organizations were offered roughly equivalent decision-making roles. That there was a place for whites both in the UDF and the greater struggle was largely undisputed among the UDF affiliates. However, no one was setting an automatic place at the table for whites. The simple presence of whites may have been an eloquent statement and a triumph in 1953, but in

1983 whites had to define a role for themselves in the anti-apartheid matrix, and were expected to act on that definition.

From the time of the government's constitutional initiative and the people's counter-thrust in creation of the UDF, there was an escalation in popular anti-apartheid initiatives and in the vigor of the state's response. The 1980s were highly fertile years, both in the dynamism of activity undertaken on many levels and intellectually as well. The white left convoked itself a number of times in the mid-1980s to deliberate its role in what appeared, to many, to be the final round in the battle against apartheid -- the beginning of the end. From the records of these deliberations one can conclude that a rough consensus had emerged among whites identifying with the UDF and embracing the call for a democratic, non-racial South Africa.

Central to these deliberations was the belief that the consensus within the white ruling "bloc" was breaking down. This bloc, which had enjoyed strength by virtue of a consensus of need and belief, was now fracturing. It was undergoing what one observer described as "the disintegration of a previously monolithic entity." This was a cause for hope.[13] The fracturing of consensus was said to be the result of a "crisis of legitimacy" in the government, a crisis caused by the government's inability to manage the multi-faceted popular revolt and exacerbated by international condemnation, loss of investor confidence, and economic crisis. This crisis was reflected in a loss of support for the governing party and the creation of new political forms to the left of the Nationalists. This fallout of the crisis suggested to white analysts on the left the primary role whites might play in exacerbating the ruling bloc's crisis to fruition, that of disorganizing:

> White democrats must tap this source [the power of the masses in struggle], and exploit the contradictions and fissures which emerge within the ruling bloc. So in playing its disorganising role, therefore, this means exacerbating the contradictions which emerge from the laager from which white democrats emerge. Our fundamental aim here is to reduce the size and the power of the enemy to its smallest and weakest point -- to isolate it -- to locate it clearly in juxtaposition to the popular democratic forces.[14]

As the 1980s progressed, there was increasing evidence that the white bloc was indeed becoming increasingly disorganized. Defections from the National Party increased and new organizations and public-interest centers formed to discuss and promote the role of whites in actively changing South African society and building toward a post-apartheid era. Whites on the left working to disorganize the white ruling bloc may have often felt, like their COD predecessors, that work with blacks in black communities was more rewarding and vital. However, unlike those COD counterparts, whites working in the national liberation movement in the 1980s found substantial evidence that the society which both they and their forbears had sought to destroy was, in fact, disintegrating. The whites working among the broader community found still greater value in their work because they had come to define their community, the white community, as unique. White analysts on the left noted that, unlike other communities, whites formed "the basis or foundation of state power and capitalist power."[15] This understanding dictated that whites would have to implement strategies in their work "completely different from a strategy in oppressed communities."[16] The strategy of disorganization was one aspect of overall white strategizing. As Raymond Suttner described it, "The entire cohesion of the enemy camp appears to have been undermined and there appears to be a general lack of confidence amongst whites, manifested in desertions from the army; emigration; adverts as to how you can get your second passport, etc."[17] The other side of the self-described white role was that of organizing, of taking that which had been freed from its old moorings in the process of disorganization and disintegration and bringing it from the "enemy camp" into the "people's camp":

> In broad terms, this means shifting people right out of the ruling bloc, and locating them concretely within the popular democratic struggle; the popular democratic organisation. It means recruiting people into our organisations, and strengthening our organisations by sound political education and analysis. It also means training people to participate and support the struggles of the oppressed and exploited majority.[18]

Disorganization and organization: These were the ideal dual roles that democratic whites defined for themselves in what many believed was the final push against the apartheid state. These definitions arose out of both expectations and experience. They were undoubtedly influenced by the larger framework of the national liberation movement and ideas held among leading ANC members as to the most useful roles whites could play in the overall struggle. They also arose from the experiences of white intellectuals who were actively involved in anti-apartheid work, liaising with or working within larger anti-apartheid structures and attempting to construct and hone a white role in the context of these experiences.

The shared vision of democratic whites was the creation of "a new nation where South Africa belongs to all who live in it, in a relation of equality."[19] The process of translating the theory of a white role into practice within the broader anti-apartheid structures of the national liberation movement raised an important question among the white participants themselves: Could one destroy an apartheid society and put a non-racial democratic society in its stead while accepting or recreating in one's political work, i.e., in one's organization, the very boundaries of race and ethnicity and class one was charged with destroying?

The Johannesburg Democratic Action Committee, the UDF, and the Cape Democrats: Rehashing the "One Congress" Question

Whites seeking engagement in anti-apartheid activities found a number of organizational outlets for their energies as the 1980s progressed. The End Conscription Campaign (ECC) focused largely on a single issue, conscription and military service. A number of liberal organizations, the Black Sash being the most prominent example and the South African Institute of Race Relations, persevered in their anti-apartheid work or, like the Five Freedoms Forum, the Institute for Democratic Alternatives, and Concerned Citizens, were founded during, the 1980s. Whites could be found in trade unions and other labor-related organizations, in social welfare, research, watchdog, and support organizations. This section focuses on whites who sought to align themselves directly with the national liberation movement and to be seen as so aligned. As a member of JODAC, the UDF, or the Cape Democrats (CD), one was aligning oneself with the recovered tradition of the ANC, the Congress Alliance, and the Freedom Charter. One was proclaiming the complete illegitimacy of the political, social, and economic status quo and was stating one's willingness to take risks in doing battle with it. Finally, one was aligning oneself with all those South Africans who shared a vision of a non-racial and democratic future. Members of JODAC, UDF, and the Cape Democrats were publicly declaring their departure from the ruling bloc and engaging in a sometimes tentative and awkward initial exploration of the possibilities of a society and culture beyond apartheid. Whether the whites involved were conscious or not of the risks inherent in that exploration, this was an undertaking requiring a great deal of courage, for one was liable to come face-to-face not only with exciting possibilities, but with complications and limitations.

The Johannesburg Democratic Action Committee: "No Is Not Enough!"

The author of a 1986 JODAC report noted that prior to JODAC's formation in August 1983 the white left in Johannesburg had been "disparate" and "divided." Whites had worked on the anti-Republic day campaign in 1981 and had given support to the Fattis and Monis strike and boycott; individuals could be found in service groups and trade unions. However the community as a whole lacked a reference point, a "defined relationship" to the progressive movement.[20] The formation of the UDF was the proximate cause of JODAC's formation. Meeting to discuss the meaning of the UDF's founding for white democrats, whites decided that they must respond to the government's plans for a whites-only election, which sought affirmation of the planned tricameral parliament and revamped executive. White activists founded JODAC to work within the white community and counter propaganda around the government's proposals and the upcoming election. JODAC appeared to fill a void for white democrats in Johannesburg, for the organization chose to affiliate to the UDF shortly after its initiation and was active throughout the 1980s, disbanding only in 1990.

At its annual meeting in 1984, a "Report on JODAC and the White Constituency" attempted to define JODAC's actual and potential constituency. Among those constituents noted were present and former JODAC members, "groups of people such as democratic academics and lawyers, the churches, journalists, the Black Sash and other progressive organisations."[21] The report also noted "an ill defined group of liberals that we believe to exist, including the left wing of the PFP [Progressive Federal Party] and others with no party affiliations."[22] The core of JODAC remained those committed activists who had sought involvement in the national liberation movement prior to JODAC's formation, many of whom would seek continued involvement after the imposition of the first State of Emergency in 1985. There was some fear that JODAC might become " 'another Congress of Democrats,' " an apparent allusion to the writer's feeling that COD had enjoyed more influence than was its due.[23] In fact, one could argue that JODAC set itself well apart from the Congress of Democrats of the 1950s in that JODAC took its role in the white community quite seriously. This is evidenced in the 1984 report to the annual meeting, where the organization is criticized for lagging behind in addressing the broader white community. The report's writer criticizes the organization for insensitivity to the varying beliefs, concerns, and degrees of commitment among JODAC's potential constituency. Specifically, the author notes that JODAC knew "little about the current developments within the white liberal community, and how they have been reacting to the increasing level of political resistance, and whether this has made them more susceptible to involvement in and support for extra-parliamentary opposition. This is a subject in need of urgent analysis."[24]

JODAC took the time to do some of this analysis. Among the constituencies JODAC members defined as approachable in the white community at the time were "Black Sash, liberals, the so-called 'Rockey Street' constituency, i.e., young punks, women, churches and the non-JODAC left."[25] Based on its research, the organization mounted a debate around political alternatives for liberals, staged a concert for the Rockey Streeters, and arranged a cultural evening with appeal to Afrikaners. By the time of the annual meeting in November 1985 the author of the annual report was able to boast that by late 1985 JODAC had broken its political isolation and was "gaining credibility for UDF politics in white Johannesburg."[26] This success was not uncomplicated, however. Focusing on its intended constituency, members of JODAC were acutely aware that, from late 1984 and escalating through 1985, the mood of the country generally was highly militant; the townships were revving at a high political pitch, and the state was meeting this resistance with force and censorship:

> As a result a debate emerged about whether we were responding adequately to township struggles. We criticized ourselves for not being able to add to the resistance and provide non-racial solidarity. However, more seriously, we were also out of touch with our community and so we consistently underestimated the extent to which the township violence and the declining economic situation were shaking their security and outlook and thus the potential to mobilise and organise whites. Without realising what it meant, we would be constantly surprised at the good attendance of meetings despite poor advertising.[27]

JODAC's membership and its constituency made various demands on the organization. Many whites participating in JODAC did so with the expectation that they would come in contact there with black comrades, or would be directed into work within the townships. Some JODAC members *were* working in the townships, and internal documents noted that government repression against the organization was primarily in response to the "valuable resources and skills" provided by JODAC members to township organizations "as a result of [our] privileged position."[28] Yet among JODAC's leadership there was also keen recognition that the role of whites in "assisting unions and township organizations [was] a diminishing one."[29] Perhaps most important, JODAC leaders understood that whites were not "best placed" to work as organizers in black communities:

> On a mass level whites couldn't do that kind of organisation, they don't speak the language, they're not there, they leave, they dominate in all sorts of ways with their resources, they don't actually assist. Unless they're very good at it. There were a few. So it's not a thing of principle, that whites must never do that, but ultimately if you want to actually do political work, and not just make yourself feel good, then you have to work in your own constituency. And you have to provide the access that you have to other people and the information, and carry over information to other people who are not going to do those things.[30]

Whites who had joined JODAC in the hope of making abiding contact with blacks found that in fact the majority of their organizing time was spent among whites. JODAC members enjoyed some sense of connectedness to the broader struggle through the outreach efforts of the national UDF leadership. The 1986 "Call to Whites" campaign initiated by the UDF was a great aid to JODAC. JODAC described the campaign as "the high point of the phase of mobilisation and expansion."[31] The theme of the campaign was an appeal to white South Africans to make a conscious decision to stay and contribute to the building of a non-racial, democratic South Africa. JODAC had interpreted its role largely as that of bringing the UDF to a frightened and insecure white community. The UDF's direct call to whites was helpful in this regard. During the campaign JODAC added five new branches to those two already in existence. However, the attention the campaign brought to JODAC's work greatly increased state repression against the organization. As a result, JODAC decentralized its leadership and changed its focus, turning from strongly UDF-based work to white area work in concert with other white organizations. JODAC played an important role in the formation of the Five Freedoms Forum (FFF), a kind of umbrella organization uniting anti-apartheid organizations in the white community.

Given the white left's definition of its role as a disorganizing and organizing force within the broader white community, its role in the formation of FFF was wholly appropriate. JODAC involved itself in a number of initiatives in the latter half of the 1980s, FFF being just one. A number of primarily white organizations came together in the Concerned Citizens (CC) to protest the declaration of the State of Emergency. The Black Sash (BS) was responsible for CC's formation, but BS and JODAC were joined by NUSAS, the Progressive Federal Party (PFP), the Detainees Parents Support Committee (DPSC), and others.[32] Having made the decision to enter CC, JODAC found that CC provided a forum in which JODAC and other progressive forces could challenge powerful liberal organizations like the PFP and lead them to more progressive stands than one might have otherwise expected.[33]

The Five Freedoms Forum, another coalition effort largely composed of white organizations, took CC's effort a step further. Growing out of the State of Emergency as had CC, the Forum brought together dozens of organizations in an attempt to enunciate a common response to government repression and to work together on issues of common concern. Organizations as diverse as JODAC and the PFP, Young Christian Students and the Lawyers for Human Rights, Women for Peace and the Anglican Board for Social Responsibility joined the initial talks. FFF was formed from these in March 1987 as a campaign-based organization.[34] The Forum's self-proclaimed ultimate goal was the attainment of a non-racial democracy in South Africa through negotiation. The FFF's task was to prepare whites for a negotiated settlement, while uniting those organizations within the Forum. The Forum participated in campaigns, including "Christmas against the Emergency" and the "Free the Children Alliance." The Forum held a number of successful conferences, including a September 1987 conference entitled "Towards Democracy: Whites in a Changing South Africa," which was attended by over 800 delegates.[35] As part of a progressive, broad-based coalition, FFF members were acutely aware of the context of isolation and fear in which whites contemplated the possibility of a common future. They were also aware that whites who approved of change but were not members of any relevant organization often felt frustrated, having no means or outlet for acting on their beliefs. The FFF publication "101 Ways to End Apartheid" was one project that attempted to speak to white fears and desire for involvement. Suggestions on the theme were solicited from whites and blacks, and they were telling. One writer stated that it was necessary to stop thinking in divisive racial and ethnic terms: "We need to think post-apartheid, act post-apartheid and live post-apartheid."[36] Other participants wrote of the plight of domestic workers and the need for white employers to pay them decent wages, treat them like adults, and aid them in their efforts at organization. A black correspondent declared that whites must "learn to speak to us" and refrain from using baby talk or barking commands. JODAC's participation in these coalitions and activities placed it squarely within the progressive white community in Johannesburg, where it could present the viewpoint of the UDF and attempt to influence liberal thought in that direction. JODAC members considered this to be valuable work, but they continued to experience a nagging sense of isolation as well.

By 1988 the JODAC leadership was noting a general despondency in the organization, coupled with a lack of direction and "alienation from black politics and the townships."[37] State repression was at its apex. The UDF had been restricted, the State of Emergency was in effect, and hundreds of activists were in detention, on trial, or serving terms in prison. JODAC documents in 1988 and 1989 suggest that JODAC members felt distanced from the core struggle and from their black, particularly African, counterparts. There was a need to find ways for white activists to meet their black counterparts and "regain a sense of what is going on in the townships."[38] Minutes from a JODAC meeting in early 1989 call on the organization to "find ways to facilitate non-racial contact. We must begin to see building non-racialism as more than just meeting with black activists. We must develop a feeling of what living in an oppressed community is all about."[39] However, the Emergency made the building of non-racialism a considerably more difficult project on the level of direct human contact. Activists were under seige from the state, making movement and congregation more difficult generally, and the anger of township youth made whites' presence in the townships a potentially dangerous proposition. JODAC proved a truly useful organization to the UDF and the national liberation struggle, as a member and a representative in a community otherwise difficult to reach. Any frustrations engendered by JODAC's relationship to the UDF and the national liberation struggle were largely a function of frustration at their inability to connect with black counterparts and to feel that they were building the non-racial future in the apartheid present.

The UDF's Cape Town Area Committees

In Johannesburg white democrats reacted to the formation of the UDF by forming JODAC. JODAC's counterparts in Cape Town chose instead to affiliate directly with the UDF, forming "Area Committees" in the largely white Cape Town suburbs of Gardens, Observatory, and Claremont. In so doing, these whites became the only formal "members" of the UDF: All other individuals were UDF members by virtue of membership in a distinct affiliate organization such as a trade union or community organization:

> At the beginning, let us note that our status in the UDF is something of an anomaly. UDF is a Front of affiliated organisations working in different constituencies. It is not a political party, and therefore makes no provision for individual membership. Area Committees are supposed to be co-ordinating structures for the branches of all UDF-affiliated organisations in a particular area to co-ordinate political work in that area. Membership of the UDF occurs through membership of an affiliate organisation....Almost from the beginning, the Area Committees were structured as individual membership *volunteer groups* of the UDF, and ever since then have functioned as such.[40]

Why did whites in Cape Town choose not to form a distinct white affiliate like JODAC? The probable primary reason was their sense of a durable tradition of non-racialism, held strongly by many Cape Town democrats. The notion of forming a racially defined organization in the service of non-racialism was undoubtedly abhorrent to some. The white Area Committees were dissolved only at the end of 1987, and between the time of their formation and their demise these unique UDF satellites attempted to bridge township struggles and suburban white disaffection.

Much like their Johannesburg counterparts, the leading members of the Area Committees defined their primary role as that of hastening the process of liberation through weakening the power of the apartheid regime. This was to be accomplished by "winning over a broad range of whites," and those activists best placed to do so were "*white democrats*."[41] This role was defined, in part, by a realization of what whites could not do. Writing of white roles outside the white areas, one Area Committee author noted there were two types of work in which white democrats could engage: that of individuals offering particular skills as needed and "the assertion of *non-racialism* through the *symbolic* importance of our presence in grassroots work."[42] The author noted that the scope for whites' participation in either was limited. As were the expectations among Area Committee members of being able to influence whites. Area Committee leaders recognized that they would not succeed where the PFP had failed in changing the hearts and minds of English-speaking government supporters.

The Area Committees enjoyed some successes. Mass meetings were attended at times by hundreds and even thousands of Capetonians. Whereas an August 1983 meeting in Observatory attended by 300 was considered a triumph, the Area Committees' inaugural effort in Sea Point was attended by 1200.[43] Who came to such public meetings? One Area Committee participant described them as an array of ex-hippies, professionals, ecologists, those who had done their national service and were disaffected by it: "UDF politics [provided] the only explanation and contact point for people."[44] The Area Committees were important conduits for information flowing from, and about, the townships. If there was such a thing as an average Area Committee member, that person probably was young (mid-twenties or less), single, and comfortable. Their reality was far removed from that of the townships. They were seeking some connection across that gulf, an authentic voice of experience and anger, and the Area Committees sought to provide it:

> I think at the least, that the Area Committees working in white areas had to continuously bring in and connect black speakers, bring in information. The flow was there, although one was continuously having to explain to people why they couldn't go and work in Guguletu. Why, if you're building grassroots structures, you didn't actually need one thousand whites with their cars and telephones and money coming in to organise: that that would actually not assist.[45]

Participation in a UDF Area Committee may have provided a young white person's best chance of connecting across the gulf of ignorance, anger, and misrepresentation that only increased with the State of Emergency. The Area Committees provided some sense "of what you were a part of, and [a sense of] acceptance of whites."[46] However, the anomalous nature of Cape Town's predominantly white Area Committees also created problems unique to them. As direct adjuncts of the UDF, rather than independent allies, the Area Committees' mission was to explain the UDF's philosophy and aims to an audience that would otherwise not have been so exposed: the liberal white community of Cape Town. The fact that these committees were predominantly white was intrinsically awkward. White Area Committee members were opened to charges of exceptional treatment, and their generally anomalous position, of which they were already acutely aware, was emphasized.[47] Members of the Area Committees may have been more concerned with their anomalous position, however, as it impinged upon their ability to act. Though chartered to work among the white community as representatives of the UDF, Area Committee members found themselves constrained at times in their attempts to build bridges with liberal counterparts for the very reason that, as members of and direct subordinates to the UDF, they were not in a position to make formal arrangements with other organizations:

> While it is recognised that as UDF Area Committees we represent nothing other than the Front itself and therefore can have no formal agreements with the PFP, the question of the political correctness or otherwise of relations with the PFP at certain points still needs to be examined. If we feel that it is necessary to enter such relationships at certain points, but that our status (as

> Area Committees rather than affiliate organisations) precludes this, then we must be prepared to reconsider the pro's and con's of changing our status. This is not simply a question of expedient name-changing to disguise a certain reality: White democrats and all activists need to be clear on whether it is necessary to have relations with the PFP or not....Our essential objective remains the same: *to win the support and where possible the involvement of PFP supporters and members in the democratic struggle.*[48]

Debate around the question of whether white democrats in Cape Town needed an independent organizational base from which to work escalated through the mid-1980s. Those in favor of the dissolution of the Area Committees and the formation of a Cape concomitant of JODAC argued that white democrats' primary task should be that of "investigating which structures best advance our work, and making these decisions on a political basis."[49] Proponents of restructuring also pointed to the upswing of interest in the committees and in the UDF generally and noted that their current status as representatives of the UDF did not always allow white democrats the flexibility necessary to work effectively in a white community that was "necessarily of a more moderate character -- especially in tone -- than the character of the broader UDF."[50] Rejecting the argument that the formation of a specifically white organization would contradict the non-racial goals of the struggle, proponents suggested that work in the white community was the most affirmative work whites could do in the name of creating a non-racial society. Interestingly, the Congress of Democrats, "an essential part of the non-racial tradition of the liberation struggle," was pointed to as evidence that the formation of a white organization would not contradict non-racialism in any sense. And therein lay the crux of the matter. The Congress of Democrats was an animal of the 1950s. Perhaps the fear on the part of those opposed to a white organization centered on the thirty intervening years: Would the formation of a white organization similar in many ways to COD prove that little had changed, that little progress had been made since the 1950s, and that South Africa, even the microcosmic South African left, was no nearer the realization of non-racialism than had been its 1950s precursor?

Those opposed to the formation of a distinct white organization felt that it was potentially compromising "to move onto the same terrain" as liberal organizations.[51] Doing so would reflect legitimacy on the PFP and other liberal groups. This, in turn, could create conflict within the UDF. Opponents of a white organization also argued that the Area Committees had played important roles organizing meetings, doing door-to-door work on various campaigns, participating in forums and workshops, distributing UDF publications, and "building progressive non-racialism in practise."[52] Opponents believed that the Area Committees were working effectively in what they considered the desirable context of UDF membership. What was not broken should not be fixed. Ultimately the proponents of dissolution of the Area Committees and formation of a distinct white organization carried the day. The Cape Democrats (CD) was launched in Cape Town on 24 April 1988, over four years after its Johannesburg and Grahamstown counterparts were chartered. The organization's inaugural declaration stated that the Cape Democrats acknowledged "the tradition of non-racial opposition which has always welcomed any person committed to democracy. We believe this places a special responsibility on those of us who are privileged by apartheid."[53] The organization's explanatory pamphlet declared that the Cape Democrats was chartered to "fill a gap in the South African political spectrum and be a political home for white democrats."[54]

In the 1950s, the evocation of the "One Congress" question was, at base, about the relationship between theory and practice, between envisioning a microcosmic non-racial culture and the exigencies of struggle which dictated that black and white would organize more effectively by organizing apart. The decision by Congress of Democrats elders not to challenge the multi-racial structure of the 1950s Alliance was neither a decision of consensus or of complete comfort. It was not obvious to all white democrats in 1958 that the greater good of building a liberation movement and achieving a non-racial democracy would best be served by maintaining racial and ethnic divisions among those working for these ends. The acquiescence of most white Capetonian activists to a separate organization in 1988 was no more an endorsement for multi-racial organization in a non-racial struggle than had been the decision of COD leadership to accept the Congress Alliance's status quo thirty years earlier. White activists, probably to greater extent than any of their black counterparts, needed to feel in a more overtly physical and geographical sense the power, the clarity, and the possibility stemming from shared experiences with African, Indian, and Coloured comrades. As the heretics of the "ruling bloc," white democrats needed to make non-racialism palpable, and they believed in turn that one of the most important contributions they could make to the national liberation movement was, simply, their presence.

Non-Racialism, Transformation, and Power

In an interview in May 1993, André Odendaal, a South African scholar and activist, describes the preceding ten to fifteen years of his life, the curve of his rise to political consciousness, activism, and historical scholarship as a period of "crossing boundaries"; he describes the trauma and personal enrichment therein.[55] Odendaal met black South Africans socially for the first time as an adult while playing cricket. A group of non-racial cricketers invited Odendaal to a post-game party:

> That's what I'm talking about, crossing boundaries. Once you've met black South Africans on equal terms, you've had supper with them, you've talked, you find them dignified and interesting, there's no way that those boundaries could lock behind me again. I couldn't go back there and lock myself away. Writing that cricket book [a work on race and cricket self-published while a student] exposed me to the other side of the fence. Before that I'd never known black people on a social level or on a level of equality.[56]

Odendaal had been a student at the University of Stellenbosch in the Cape, had left for Cambridge, and then had returned to South Africa and enrolled in the University of South Africa (UNISA), where he attempted to fit back into the South African sport niche. Radicalized by then, he could not stomach the chauvinism and bigotry of the South African sport world. Odendaal first turned to liberal circles for advice on becoming involved politically. Frustrated there, he decided to go to a UDF township meeting in January 1984, a frightening venture, as he had no idea what to expect upon reaching the township church:

> I walked into the church, and within ten minutes I knew this is where I belonged. What people were saying first of all, and secondly the kind of energy. There was a feeling of strength in that room, of ordinary people as well. I mean, you're sharing a song sheet with a woman who is a domestic worker. It was so safe in many ways, and so exciting and energetic in other ways, and I saw here people are singing 20 kilometres from the Union Buildings, in Pretoria, [and] white South Africans didn't know where they were. From then on, I was UDF, I went to meetings, I started making contact with people.[57]

The power of such moments is recounted by many white South Africans who were brought up in segregated neighborhoods and schools, reading papers and listening to other media that accepted, if not propounded, the apartheid status quo. One conscription resister noted that his normal white middle-class upbringing included hours playing with toy soldiers, reading war books, and looking forward to the time when he would enter the army.[58] Entering an African church or a similar venue in a township often proved a signal moment in these lives, especially so because these were generally moments of affirmation. Whites who ventured into black areas, particularly for events of political significance, were often treated as heroes. As one activist put it, "it's much nicer to go to a funeral where people pat you on the back and say, 'hey, comrade' than to knock on doors in Newlands and people kick you out if their dog hasn't bitten you before you get to that door."[59] Whites took a great deal away from these encounters because they felt accepted and at the same time believed that they were building something positive, a future for whites in South Africa, by their presence in the townships or at non-racial events. A Five Freedoms Forum author, writing of IDASA's (Institute for a Democratic

Alternative for South Africa) mission, noted that the organization was born to encourage and establish contact between "black and white South Africans who have been kept apart for so long that they have lost touch with each other's humanity."[60] White democrats needed desperately to know that they were not so isolated, so jaundiced and altered by the culture from which they came that it had become impossible to make elemental human connections with their black compatriots. A white recounts in an Area Committee publication his venture to a township funeral in Guguletu:

> A taxi-driver stops next to a white journalist and says, "Hey Whitey, aren't you scared?" The journalist replies, "Why should I be scared, I am with my people." The solidly packed taxi erupts with a throaty roar of "Amandla!" and "Viva Comrade." The journalist has tears in his eyes.[61]

The author, describing the scene, finds in it a vision of a future of camaraderie without bitterness and "an overwhelming feeling of belonging. A feeling that this is what it could be like if only we could have our freedom now, blacks and whites together."[62]

But what was the relationship between these moments of camaraderie and closeness and felt shared purpose and the realization of a non-racial, democratic, and just South Africa? Did non-racialism remain a live construct *because* it was lived in moments such as those described above? Or was it alive because it was embraced as an unshakable principle by so many South Africans independent of then-present realities, strictures, and complications? Were there attempts to define non-racialism? Or did its definition arise solely as an outgrowth of lived experience, if at all?

Mike Roussos, a white trade unionist working in non-racial unions, suggested that South Africans would not overcome the real differences between them by wishing those differences away. South Africans would have to recognize the problems between them and would have to work at overcoming these problems through the process of political struggle, what he described as "an active non-racialism -- working to overcome the problems that racism has caused -- not an assumed non-racialism which just says we [are] all equal participants in the struggle."[63] One might suggest that political struggle would be but one means of recognizing and treating with the differences, political, economic, cultural, and social, that were the diverse raw materials from which either a non-racial culture would be woven or a culture in transition would fracture. This is to suggest that the arenas in which one worked to foment non-racialism would ultimately define the nature of that non-racialism. Non-racialism lived only at the political level might remain a non-racialism of elites, a relatively brittle idea with little resonance or depth in any cultural sense. A non-racialism contested, and defined, in and through other venues such as educational systems, cultural and sporting bodies, labor associations, or artists and writers organizations, would provide a foothold for such ideas at a grassroots level, and so imbue the concept with anchors among the broader population: an organic process. The problem with passive non-racialism, from the perspective of individuals like Mike Roussos and other leading white activists, was that it did not seek to recognize and face the deep conflicts generated by the transformation from an apartheid to a democratic society, and so could not begin to grapple with them. That is why, Roussos suggested, "non-racialism" as a kind of label accepted automatically by the majority of anti-apartheid activists was so potentially dangerous. Roussos commented that non-racialism should be considered more a process than a product, not something that is "built" by the number of visits to a township or participation in marches, but grows constantly, through a hundred different types of activities and interactions. Principally, non-racialism is built through honest communication which would foster anger in the process of allaying it, create discomfort before dispelling it, and lead South Africans -- all South Africans, white South Africans -- to question deeply their understandings of themselves before allowing them the solidity of a new self-identification.

While the goal of achieving a non-racial society was largely lived on the political plane throughout the 1980s, the individual experience of non-racialism remained highly subjective and psychological; therefore it could be as variable as the number of individuals experiencing it. Whites who embraced non-racialism had very personal understandings of and associations with it beyond the rhetorical non-racialism of organizational declarations and protest speeches. Non-racialism could be the moment the idea of a "common humanity" was first lived by virtue of an individual's inclusion. Non-racialism could be an intensely personal moment in which one realized that there was a choice between joining the Defense Force and choosing to reconstruct oneself in prison. Non-racialism could be a person's presence in the townships in solidarity with black comrades. But embracing a vision of a just society and attempting to share those possibilities and other lessons learned with other whites was no less an act of non-racialism. Generally, an individual first built an idea of what non-racialism was, within, and for, him- or herself. This was a healthily selfish act, and from it stemmed the will and the sense of purpose necessary to find a place in a struggle that one could easily have defined oneself out of. Ultimately, whether the person was white or black or brown, embracing non-racialism was acknowledging the bottom line, the belief that a new South Africa whose citizens were not trying to construct a land of justice, peace, and, at minimum, toleration, would not be a society in which one could afford to live. Ergo, the danger in a white Cape Youth Congress author's statement that "The importance attached to the color of one's skin should be the same as that attached to, say, the colour of one's eyes, i.e., not at all."[64] Wishing that South Africans could simply move beyond their prejudices was counterproductive among those working for real change. The fact remained that South Africans were going to have to face, in many greatly uncomfortable moments and ways, the import of their prejudices, fears, and expectations. To wish away these moments of confrontation yet-to-come, no matter how well-intentioned, was to exhibit a keen lack of understanding of the process of building non-racialism.

Skating on the Interregnum: 1990 and Beyond

Watching Nelson Mandela walk out of Pollsmoor Prison in February 1990 was an experience that would have been almost inconceivable even months beforehand. This was the realization of reveries, a moment that had been envisioned countless times on podiums, in song, and on the printed page: the end of Apartheid. Suddenly the ANC, the South African Communist Party, and open political activity generally were unbanned. Exiles, some who had been out of the country for more than a quarter century, returned. The minds of all South Africans turned to questions of transition, some with great expectations, others with fear: When would the reins of power shift from the white minority to the black majority, and real change occur? "Real change" could only be understood as a shift in the balance of political and economic power, a matter of tremendous stakes and great complexity. Where talk of a "non-racial, democratic society" had slipped easily off the tongue in the 1980s, there was a greater necessity to define terms in the wake of the events of February 1990.

This interregnum, between the known world of repression, illegality, and exile and the uncharted territory of governance, has proven an extremely difficult time physically and psychically for people who actively fought the apartheid system and for all those who have borne the brunt of it. The certainties of struggle have given way to the uncertainties of a victory that is not absolutely trusted, its topography in the process of negotiation, largely removed from the streets and from view. The collectivity of struggle appears to have given way to a degree of atomization. The goals that held a truly diverse array of activists together, the unbanning of the people's organizations, the release of leaders, the end of the apartheid system, have been realized. The violence of the early 1990s -- what many people including activists describe, perhaps ironically, as the destruction of the social fabric; the realities of the divisions in South African society; and the very real, tremendous challenge of attempting to reorganize the ANC as a *political* body in a *political* contest -- have left many activists unsure of the outcome of their labors, and unsure of what place they may find in the South African future.

The vision of non-racialism in the Freedom Charter, renewed and shared in the 1980s, may also prove a victim of the interregnum and the economic and political realities of the South African future, realities that are abiding rather than new but that were obscured in the name of national liberation. One must return to the roots of the national liberation movement, African nationalism, the proponents of which had spoken forcefully of liberation in terms of relations of power and ownership in the 1940s. The ANC was reborn in the 1940s as an African liberation movement. From the 1950s, when the ANC's leadership had presided over the formation of the Congress Alliance, through the formation of a non-racial fighting force and exile in the 1960s, the renaissance of an internal ANC presence in the 1970s, and the ascendance of the organization within South Africa and internationally in the 1980s, the tension between the ANC's primary mission as an *African* liberation movement and its self-identification as a movement of and for all freedom-loving South Africans was never resolved. Thus, the rhetorical vision of "non-racialism" as presented by the ANC and its allies may have proven an effective and noble means of organizing a liberation movement in a highly diverse national setting, creating a working unity among communities that had been actively, sometimes violently, antagonistic towards one another in the past. But one might argue that the non-racialism of the national liberation movement had masked the very real tensions there between the goals of a non-racial, democratic society and the particular goals of the African majority: the realization of full citizenship and with it economic and political redress.

Delaying discussion of the realities attendant upon these tensions would only be delaying the inevitable. This is not to suggest that South Africa would, in the final analysis, be other than a non-racial society. But one must define one's terms. As one white activist described it, the non-racialism of the South African future might well be a "non-racialism of pragmatics," a non-racialism based in the reality that millions of whites would remain South African citizens who must be accommodated and who could contribute to nation-building by virtue of their skills.[65]

When asked by the author what a non-racial South African culture would look like, most of those interviewed in 1993 had no ready answer to the question. Individuals were preoccupied with the economic realities of the country, the immense problem of meeting the basic human needs and expectations of tens of millions of impoverished people, and how these problems would impinge on the creation of a stable and democratic society. There was a shared sense that non-racialism was no longer a matter of central concern, debate, and discussion. A number of individuals noted that the majority of the "young lions" in the townships were growing up without any contact with whites. Others noted that non-racialism conflicted with real struggles for the spoils of victory: Africanism, the doctrine claiming South Africa as an African country and proclaiming the primacy of the African people therein, was in the ascendant as it related to the dispensation of jobs and resources.[66] If, as described above, the non-racialism of many African activists in the 1980s onward was more an article of faith than a product of experience, the atomization among activists from the later 1980s and the sense of rising Africanism among Africans from the early 1990s would not appear to bode well for a non-racial future. At a time when South Africans of diverse socio-cultural and economic backgrounds need to speak to one another, and create abiding channels for communication and interaction, the tensions within and between communities, engendered by the changes and uncertainties of this period, may tend to depress or defeat such efforts.

Ultimately, there is no reason why "non-racialism," what it has meant or will mean, should be less contested than any other aspect of South African society-in-transition. Non-racialism was described earlier as a process. If that is indeed so, then those who hold dear the possibility of realizing a society that draws upon all South African cultures to create a uniquely South African culture, a culture transcending description by the language of division, may yet realize that vision. There are, however, no guarantees. First, the people of South Africa must come to terms with the great density of their past, the ambiguities and vagaries of their present, and the various consequences, welcome and unwelcome, inherent in their hopes for the future.

Notes

1. For a discussion of the rise of SASO and SASO's break with NUSAS, see Gail Gerhart, *Black Power in South Africa: The Evolution of an Ideology* (Berkeley: University of California Press, 1979), 260-265.

2. Interview, Mike Roussos, by Julie Frederikse, in the South African History Archive, Johannesburg.

3. For a discussion of Richard Turner's background and his politico-economic analysis, see Richard Turner, *The Eye of the Needle: Toward Participatory Democracy in South Africa* (Maryknoll, NY: Orbis Books, 1978). Richard Turner was assassinated by a person or persons unknown in January 1978.

4. Graeme Bloch, NUSAS Congress speeches, 1981, Peter Williams Collection, Archives of the Mayibuye Center, University of the Western Cape.

5. Julie Frederikse, *The Unbreakable Thread: Non-Racialism in South Africa* (Bloomington: Indiana University Press, 1990), 139.

6. Interview, Alec Irwin by Julie Frederikse, in the South African History Archive, Johannesburg.

7. Ibid.

8. Jeremy Cronin, interview by Julie Frederikse, in the South African History Archive, Johannesburg.

9. Ibid.

10. Rob Davies, Dan O'Meara and Sipho Dlamini, *The Struggle for South Africa: A Reference Guide to Movements, Organizations and Institutions, Volume Two* (London: Zed Press, 1984), 370-371.

11. Mike Evans, NUSAS Congress speeches, 1981, Peter Williams Collection, Archives of the Mayibuye Center, University of the Western Cape.

12. United Democratic Front, "Declaration," 20 August 1983, Rocklands Civic Center, Mitchells Plain, Western Cape.

13. Article, "Getting Our Act Together," Michael Evans, December 1987, in Cape Democrats file, South African History Archive, Johannesburg.

14. Tom Waspe, interview by Julie Frederikse, July 1985, in the South African History Archive, Johannesburg.

15. "Proposal for a UDF Campaign Around a Call to Whites by the ANC," 30 January 1990, n.a.

16. Ibid.

17. Raymond Suttner, speech given at a conference on the role of white democrats, Johannesburg, January 1986.

18.Tom Waspe, interview by Julie Frederikse, July 1985, in the South African History Archive, Johannesburg.

19. Raymond Suttner, speech given at a conference on the role of white democrats, Johannesburg, January 1986.

20. Johannesburg Democratic Action Committee, "Report for national workshop," January 1986, in the South African History Archive, Johannesburg.

21. Johannesburg Democratic Action Committee, "Report of the JODAC AGM, 11/11/84, Report on Political Education and Debate," JODAC File, South African History Archive, Johannesburg.

22. Ibid.

23. Johannesburg Democratic Action Committee, "Report for national workshop," January 1986, in the South African History Archive, Johannesburg.

24. Johannesburg Democratic Action Committee, "Report of the JODAC AGM, 11/11/84, Report on Political Education and Debate," JODAC File, South African History Archive, Johannesburg.

25. Johannesburg Democratic Action Committee, "Report for national workshop," January 1986, in the South African History Archive, Johannesburg.

26. Johannesburg Democratic Action Committee, "Report for JODAC General Meeting, 3 November 1985."

27. Johannesburg Democratic Action Committee, "Report for national workshop," January 1986, in the South African History Archive, Johannesburg.

28. Johannesburg Democratic Action Committee, "Report for JODAC General Meeting, 3 November 1985."

29. Johannesburg Democratic Action Committee, "The role of White Democrats in the National Democratic Struggle," paper delivered by Tom Waspe, n.d.

30. Interview, Graeme Bloch, Johannesburg, 7 May 1993.

31. Johannesburg Democratic Action Committee, "Report for national workshop," January 1986, in the South African History Archive, Johannesburg.

32. Johannesburg Democratic Action Committee, "Report for national workshop," January 1986, in the South African History Archive, Johannesburg.

33. United Democratic Front, "Broadening the Base: Our Work Among Whites," Cape Town, 1986, Mayibuye Center, University of the Western Cape.

34. David Webster, "The Five Freedoms Forum," n.d.

35. Ibid.

36. Five Freedoms Forum, "101 Ways to End Apartheid," n.d.

37. Johannesburg Democratic Action Committee, "JODAC Executive Report to the AGM, 21 May 1989," JODAC File, South African History Archive.

38. Johannesburg Democratic Action Committee, " National Initiative, 1988."

39. Johannesburg Democratic Action Committee, "Minutes, N.C.C., February/March 1989."

40. United Democratic Front, "Broadening the Base: Our Work Among Whites," Cape Town, 1986, Mayibuye Center, University of the Western Cape.

41. Ibid.

42. Ibid.

43. Ibid.

44. Interview, André Odendaal, 19 May 1993, Cape Town.

45. Interview, Graeme Bloch, 7 May 1993, Johannesburg.

46. Interview, André Odendaal, 19 May 1993, Cape Town.

47. "Low key Cape Democrats launch seen as significant move for whites," *Southscan: A bulletin of Southern African affairs*, Vol. 2, No. 32 (27 April 1988) 1.

48. United Democratic Front, "Broadening the Base: Our Work Among Whites," Cape Town, 1986, Mayibuye Center, University of the Western Cape.

49. Ibid.

50. Ibid.

51. "The organisation of whites in the Western Cape," n.a., n.d., UDF File, South African History Archive, Johannesburg.

52. Ibid.

53. "Declaration," Cape Democrats, n.a., n.d.

54. "What is Cape Democrats?," Cape Democrats, n.a., n.d.

55. Interview, André Odendaal, Cape Town, 19 May 1993.

56. Ibid.

57. Ibid.

58. Press statement, George Galanakis, 1989. Galanakis was one of 771 white South African men who announced that they would not participate or would cease participation in the South African Defence Force's conscription program.

59. Interview, Graeme Bloch, Johannesburg, 7 May 1993.

60. Five Freedoms Forum, "101 Ways to end apartheid," n.a., n.d.

61. " 'Hey Whitey, Aren't You Scared?': A White South African at a township funeral," *Up Front: A publication of Claremont & Observatory UDF*, No. 2 (October 1985).

62. Ibid.

63. Mike Roussos, interviewed by Julie Frederikse, in the South African History Archive, Johannesburg.

64. Cape Youth Congress, Gardens branch, Information Sheet #2, 1987.

65. Interview, Joanne Yawitch, 9 May 1993, Johannesburg.

66. Interview, Graeme Bloch, 7 May 1993, Johannesburg.

List of Acronyms

AAC	All-African Convention
AES	Army Education Service
ANC	African National Congress
ANCWL	African National Congress Women's League
APO	African People's Organisation
ARM	African Resistance Movement
AZASO	Azanian Students' Organisation
BS	Black Sash
CC	Concerned Citizens
CD	Cape Democrats
COD	Congress of Democrats
COSAS	Congress of South African Students
CPC	Cape Provincial Council
CPC	Coloured People's Organisation
CPSA	Communist Party of South Africa
DPSC	Detainees' Parents Support Committee
FFF	Five Freedoms Forum
FOSATU	Federation of South African Trade Unions
FRAC	Franchise Action Council
FSAW	Federation of South African Women
IDAMF	Interdenominational African Ministers Federation
IDASA	Institute for Democratic Alternatives for South Africa
ISL	International Socialist League
JODAC	Johannesburg Democratic Action Committee
JWC	Jewish Workers Club
LAR	League of African Rights
LLY	Labour League of Youth
MK	Umkhonto we Sizwe
NAC	National Action Council
NCC	National Consultative Committee
NCL	National Committee for Liberation
NEC	National Executive Committee
NEUF	Non-European United Front

NEUM	Non-European Unity Movement
NIC	Natal Indian Congress
NP	National Party
NUSAS	National Union of South African Students
PAC	Pan-Africanist Congress
PFP	Progressive Federal Party
SABRA	South African Bureau of Racial Affairs
SACOD	South African Congress of Democrats
SACP	South African Communist Party
SACPO	South African Coloured People's Organisation
SACTU	South African Congress of Trade Unions
SAIC	South African Indian Congress
SAIRR	South African Institute of Race Relations
SALP	South African Labour Party
SAT&LC	South African Trades and Labour Council
SLA	Students' Liberal Association
TIC	Transvaal Indian Congress
UDF	United Democratic Front
UNISA	University of South Africa
UP	United Party
WAAF	Women's Auxiliary Air Force

References

Adler, T. "History of Jewish Workers Clubs," in *Papers Presented at the African Studies Seminar at the University of the Witwatersrand, Johannesburg, During 1977*, 1-66. Johannesburg: African Studies Institute, University of the Witwatersrand, 1977.

Arkin, Marcus, ed. *South African Jewry: A Contemporary Survey*. Cape Town: Oxford University Press, 1984.

Benson, Mary. "A True Afrikaner." *Granta* 19 (Summer 1986): 198-233.

Blaut, James. *The National Question: Decolonising the Theory of Nationalism*. London: Zed Books, 1988.

Brooks, Alan K. "From Class Struggle to National Liberation: The Communist Party of South Africa, 1940-1950." M.A. diss., University of Sussex, 1967.

Bundy, Colin. " 'Around Which Corner': Revolutionary theory and contemporary South Africa," *Transformation* 8 (1989).

Bunting, Brian. *Moses Kotane: South African Revolutionary*. London: Inkululeko Publications, 1975.

Burman, A. "The South African Congress of Democrats: 1953-1962." B.A. thesis (Hons), University of Cape Town, 1981.

Carter, Gwendolen. *The Politics of Inequality: South Africa Since 1948*. New York: Frederick A. Praeger, 1958.

Clingman, Stephen R. *The Novels of Nadine Gordimer: History from the Inside*. Johannesburg: Ravan, 1986.

Elphick, Richard. "Mission Christianity and Interwar Liberalism," in Jeffrey Butler, Richard Elphick, and David Welsh, eds. *Democratic Liberalism in South Africa: Its History and Prospect*. Middletown, CT: Wesleyan University Press, 1987.

Everatt, David. " 'Frankly Frightened': The Liberal Party and the Congress of the People," unpublished paper presented to the "South Africa in the 1950s" conference, Queen Elizabeth House, Oxford University, 1987.

________. "Alliance Politics of a Special Type: The Roots of the ANC/SACP Alliance, 1950-1954." *Journal of Southern African Studies*, Vol. 18, No. 1 (March 1991).

Fridjohn, Michael. "The Torch Commando and the Politics of White Opposition: South Africa 1951-1953." In *Papers Presented at the African Studies Seminar, University of the Witwatersrand*. Johannesburg: Institute of African Studies, 1977.

Forman, Lionel and Sachs, E.S. *The South African Treason Trial*. New York: Monthly Review Press, 1958.

Frederikse, Julie. *The Unbreakable Thread: Non-Racialism in South Africa*. Bloomington: Indiana University Press, 1990.

Gerhart, Gail. *Black Power in South Africa: The Evolution of an Ideology*. Berkeley: University of California Press, 1978.

Gerhart, Gail and Karis, Thomas. *From Protest to Challenge: A Documentary History of African Politics in South Africa, 1882-1964*, Vol. 4, *Political Profiles, 1882-1964*. Stanford: Hoover Institution Press, 1964.

Goldin, Ian. "The reconstitution of Coloured identity in the Western Cape," in Shula Marks and Stanley Trapido, *The Politics of Race, Class, and Nationalism in Twentieth Century South Africa*. London: Longman, 1987.

Gordimer, Nadine. *A World of Strangers*. Harmondsworth: Penguin Books, 1984.

Hirson, Baruch. "Bukharin, Bunting, and the *Native Republic* Slogan." *Searchlight South Africa*, Vol. 1, No. 3 (July 1989).

Huddleston, Trevor. *Naught For Your Comfort*. London: Collins, 1956.

Hudson, Peter. "Images of the Future and Strategies in the Present: The Freedom Charter and the South African Left," in *South Africa Review 3*. Johannesburg: Ravan, 1985.

Irvine, Douglas. "The Liberal Party, 1953-1968," in Butler, Elphick, and Welsh. *Democratic Liberalism in South Africa: Its History and Prospect*. Middletown, CT: Wesleyan University Press, 1987.

Joseph, Helen. *If This Be Treason*. London: André Deutsch, 1963.

_________. *Side By Side: The Autobiography of Helen Joseph*. London: Zed Books, 1986.

Karis, Thomas. *From Protest to Challenge: A Documentary History of African Politics in South Africa, 1882-1964*, Vol. 2, *Hope and Challenge, 1935-1952*. Stanford: Hoover Institution Press, 1977.

Karis, Thomas and Gerhart, Gail. *From Protest to Challenge: A Documentary History of African Politics in South Africa, 1882-1964*, Vol. 3, *Challenge and Violence, 1953-1964*. Stanford: Hoover Institution Press, 1972.

Lewin, Hugh. *Bandiet: Seven Years in a South African Prison*. London: Heinemann, 1974.

Lodge, Tom. *Black Politics in South Africa Since 1945*. London: Longman, 1983.

Mandela, Nelson. *The Struggle Is My Life*. London: International Defence and Aid Fund for Southern Africa, 1986.

Mitchison, Naomi. *A Life for Africa: The Story of Bram Fischer*. London: Merlin Press, 1973.

Munger, Edwin. *African Field Reports 1952-1961*. Cape Town: C. Struik, 1961.

Nkosi, Lewis. *Home and Exile*. London: Longman, 1983.

Rich, Paul. *White Power and the Liberal Conscience*. Manchester: Manchester University Press, 1984.

Robertson, Janet. *Liberalism in South Africa, 1948-1963*. Oxford: Clarendon Press, 1971.

Roux, Edward. *Time Longer Than Rope: The Black Man's Struggle for Freedom in South Africa*. Madison: University of Wisconsin Press, 1948.

Roux, Edward and Roux, Win. *Rebel Pity*. London: Rex Collings, 1970.

Sachs, E.S. *Rebel's Daughters*. Alva: Robert Cunningham & Sons, Ltd., 1957.

Sampson, Anthony. *The Treason Cage: The Opposition on Trial in South Africa*. London: Heinemann, 1958.

Saron, Gustav. "The Making of South African Jewry," in Leon Feldberg, ed., *South African Jewry*. Johannesburg: Fieldhill Publishing Co., Ltd., 1965.

Saron, Gustav and Hotz, Louis. *The Jews in South Africa: A History*. Cape Town: Oxford University Press, 1955.

Segal, Ronald. *Into Exile*. London: Jonathan Cape, 1963.

Shimoni, Gideon. *Jews and Zionism: The South African Experience (1910-1967)*. Cape Town: Oxford University Press, 1980.

Simons, Jack and Simons, Ray. *Class and Colour in South Africa, 1850-1950*. London: International Defence and Aid Fund for Southern Africa, 1983.

South African Communist Party. *South African Communists Speak, 1915-1980*. London: Inkululeko Publications, 1981.

Suttner, Raymond and Cronin, Jeremy. *30 Years of the Freedom Charter*. Johannesburg: Ravan, 1985.

Van Diepen, Maria, ed. *The National Question in South Africa*. London: Zed Press, 1988.

Walshe, Peter. *The Rise of African Nationalism in South Africa: The African National Congress, 1912-1952*. Berkeley: University of California Press, 1971.

Index

Africanism, 5-6, 27
  nature of, 28-29
  psychology of, 26, 183
Africanists, 26
  ANC, breakaway from, 163-164
  and COD, 164, 184
  and Congress Alliance, 183-185
  and Freedom Charter, 179, 182-184
African National Congress (ANC), 1, 15, 33, 65, 244, 247, 248, 249, 254, 266
  and Africanism, 44, 184, 185-187, 216
  aims, 215, 216
  All-In Conference, 230
  banning of, 225, 227, 229
  building united front, 197, 199
  call to whites (1950s), 47
  and COD formation, 68, 69-70, 145
  COD influence on, 116
  COD, relations with, 139, 180, 211
  COD, tensions between, 212-213, 216-218
  and Congress Alliance, 215, 267
  in exile, 245
  Kabwe conference (1985), 244
  and liberalism, 27
  and Liberal Party, 164, 207, 211
  and liberals, 18, 19, 200
  Morogoro conference (1969), 244
  and Natal Indian Congress (NIC), 163
  and non-racialism, 216, 244-245, 248
  and One Congress question, 198, 212-213, 216, 217
  organizing whites, 54
  and PAC, 6
  post-banning, 245
  and SAIC, 44
  sabotage activity, 226, 231
  supporting COD, 147
  Treason Trial, 200
  whites in (1970s), 248
African National Congress Youth League (ANCYL), 19, 26, 28, 29, 43, 46, 54, 157, 214
  and COD formation, 70
  and CPSA, 30
  Programme of Action, 29, 30, 164, 175, 184, 215
African National Congress Women's League (ANCWL), 152
African People's Organisation (APO), 16, 45
African Resistance Movement. *See* NCL
Afrikaners
  and ANC, 205, 206
  as radicals, 99, 101, 113
Alexander, Ray, 15, 17
  in FSAW, 152
All-African Convention, 19, 43
All-African People's Conference. *See* COD, Pan-Africanism
All-In Conference. *See* ANC

Anderson, Eleanor, 233
Anderson, Joan, 96
Andrews, Bill, 16, 40
Anti-Coloured Affairs Department (Anti-CAD), 43
Anti-Fascist League, 40
Anti-Pass Campaign (1943), 43
Area Committees, 258-262
  as anomalous, 259, 260
  branches, 259
  composition, 259-260
  controversy surrounding, 261
  dissolution, 259
  meetings, 259
  and non-racialism, 259
  and One Congress question, 261-262
  and PFP, 260-261
  role of, 259, 260
Arenstein, Jacqueline, 135(n14)
Arenstein, Rowley, 135(n14)
Army Education Service (AES), 54, 58, 106-107
Azanian Students Organisation (AZASO), 249

Bach, Lazar, 17
Ballinger, Margaret, 18, 19, 39, 166
Ballinger, William, 18, 19
Barenblatt, Yetta, 156
Barsel, Ester, 97, 155
Barsel, Hymie, 92-93
Basner, Hyman, 43, 46, 47
Benson, Mary, 100, 123
Berman, Monty, 235
Berman, Myrtle, 235
Bernstein, Hilda. *See* Watts, Hilda
Bernstein, Rusty, 25
  in COD, 73, 74
  Congress Alliance, socializing in, 149-150, 190
  early life, 93-94
  in MK, 233
  politicization, 94
  and Treason Trial, 190
Beyleveld, Piet, 173
  in COD, 73, 102, 202, 215
  in SACP, 102, 203
  in SACTU, 102, 203
  in Springbok Legion, 102
Biko, Steve, 246
Birobidzhan, 86
Black Consciousness Movement, 6, 248
  youth moving out of, 249, 250
Black Sash (BS), 201, 202, 209, 254
  and JODAC, 255
  and CC, 257
Blackshirts, 92
Boshoff, Franz, 35-36
British Empire Service League. *See* Springbok Legion
Brookes, Edgar, 7, 19
Brooks, Alan Keith, 238
Brown, Peter, 208, 209, 210
Bundists. *See* Jews
Bunting, Rebecca, 15
Bunting, Sonia, 173
Bunting, S.P., 12, 13, 14, 105
  and "Native Republic" thesis, 15, 16, 17
Butcher, Mary, 66, 67

Cachalia, Maulvi, 67, 156
Cachalia, Yusuf, 46, 68, 156, 166
Campaign for Right and Justice, 43
Campaign for the Defiance of Unjust Laws. *See* Defiance Campaign
Cape Democratic League. *See* COD
Cape Democrats (CD), 261-262
Cape Provincial Council (CPC), 34
Cape Town Liberal Group. *See* Liberal Party
Cape Youth Congress, 265
Carneson, Fred, 25, 55, 58, 59, 105
Carneson, Sara, 25

Christian Institute, 249
Cohen, Percy, 66
Coloured People's Congress. *See* SACPO
Coloureds, voting rights, 61
Colraine, Daniel, 14
Comintern, 15
and CPSA, 14, 16, 20
Communist International. *See* Comintern
Communist Party of South Africa (CPSA), 4, 5, 12, 18, 46, 54, 71, 75, 214, 225, 247
  African leadership, 14, 25
  African membership, 14
  and African nationalism, 24, 31, 48(n19)
  and ANC, 27
  and ANCYL, 32
  and black working class, 31
  and disbanding, 49(n23), 53
  electoral activity, whites, 34-42
  and Jewish Workers Clubs, 86
  and League of African Rights, 16
  membership, WWII period, 33-34
  and national convention idea, 45
  national liberation, shift to focus on, 32-33
  and National Question, 23, 31-3, 41, 47, 48(n1)
  "Native Republic" thesis, 14-18, 20, 31
  1922 miners' strike, 13
  non-racialism in, 20, 26-27, 34, 35
  and SALP, 14
  and Soviet Union, 41
  whites as dominating, 31
  whites as leaders, 24, 25
  whites in, 24, 25, 34
  and white working class, 24, 31
  youth involvement, 92, 93, 95, 97, 102, 103
Concerned Citizens (CC), 254, 257
Conco, W.Z., 171
Congress Alliance, 2, 74, 107, 211, 217, 225, 234, 244, 248, 251
  ANC leadership of, 81
  in the Cape, 119
  formation, 72
  form of organization, 115
  legacy of, 248, 250, 254
  and Liberal Party, 8
  multi-racial activity prior to formation of, 30, 42-47, 64-65
  National Consultative Committee (NCC), 155
  and One Congress question, 198
  socializing within, 139, 140-145, 149-151, 190-191, 213
  and State of Emergency (1960), 228
  uniqueness, sense of, 219
  white participants, affluence of, 155-156
See also Congress of the People
Congress of Democrats, 72, 75, 82, 151, 252, 261
  and ANC, 115, 116, 118, 120, 169-170, 205, 220, 245
  black community, work in, 120, 122, 123, 133, 146-147, 148, 153, 168-169, 171, 210
  branches, 118, 119, 168
  and broad front organization, 202, 203, 204-205, 206, 219
  Cape Democratic League, 73
  communists in, 73, 116, 202
  and Congress Alliance, 115, 117, 122, 146, 217, 218, 220
  and domestic workers, 155-156
  Durban, 120
  education, internal, 120, 135(n16)
  electoral activity, 202, 207
  formation, 68-70, 72, 115
  Freedom Charter, role in, 169, 180-181, 192

impact, sense of, 203
inaugural conference, 72-73, 79(n68)
internal divisions, 201, 203, 204
internationalism, 73
Johannesburg, 118
Johannesburg Congress of Democrats, 73, 74
and MK, 225
leadership, 73, 116
and Liberal Party, 8, 117, 202, 204, 205, 206, 207, 208, 217, 220
Marxists in, 121
membership, 72, 134, 202
Million Signatures Campaign, 187-188
and One Congress question, 198, 211, 212-218
and Pan-Africanism, 198, 218-220
as political educator, 121-124, 135(n20)
publications, 121, 147, 157, 229, 230
recruitment, 117, 118, 202
role, self-defined, 74, 145
and SACP, 133
and SAIC, 115
self-criticism, 119, 189
and Sharpeville massacre, 228-229
and Springbok Legion, 68, 72
and State of Emergency, 228-229
structure, 117
and Treason Trial, 197
and United Party, 117
and Universal Declaration of Human Rights, 73
in Western Cape, 119, 188
white community, work in, 118, 120, 133, 168 187-188, 189
and working class, 79(n70), 204
The World We Live In, 120-124, 135(n20)
youth in, 118, 214-215
See also Congress of the People
Congress of South African Students (COSAS), 249, 250
Congress of South African Trade Unions (COSATU), 244, 247
Congress of the People, 2, 46, 120, 173, 243
and Africanists, 163
aims, 162
COD role in, 161, 164, 167-169, 171, 172
Congress Alliance affiliates, unifying, 170-171
fundraising for, 171-172
hijacking, fears of, 171
and Indian community, 162
and Liberal Party, 162, 165
National Action Council (NAC), 164, 169, 171, 175, 179
National Consultative Committee (NCC), 187
origins, 161
representation at, 163
and Sharpeville massacre, 228
significance of, 174-175
and State of Emergency, 228
Western Cape, activity in, 169-172
Cronin, Jeremy, 248, 249, 250

Dadoo, Yusuf, 25, 45, 156, 199
Daniels, Eddie, 235, 236
Defend Free Speech Convention, 65
Defiance Campaign, 65, 74, 162
and ANC, 65
impact of, 65
promoting multi-racialism, 66
and Springbok Legion, 68
whites, impact on, 65
whites, role in, 54, 66, 67, 69
Detainees' Parents Support Committee (DPSC), 257
Discussion clubs, 118, 124-133

and indigenous bourgeoisie, 132
and National Question, 130-133
and Soviet theorists, 130
and white role, 132
and working class, 131
See also Johannesburg Discussion Club
Doctors' Pact. *See* South African Indian Congress
Duncan, Patrick, 66, 67, 208
du Plessis, Danie, 25, 129
du Toit, Bettie, 3, 102-105, 158n
in Defiance Campaign, 66-67

Eisenstein, Raymond, 236
End Conscription Campaign (ECC), 254, 263

Fattis and Monis, 250, 254
Federal Party, 209
Federation of Non-European Trade Unions (FNETU), 49(n41)
Federation of South African Trade Unions, 247
Federation of South African Women (FSAW), 107
aims, 154
founding meeting, 151-152
and Freedom Charter, 153-154, 177-178
marches, 152
members, organizational, 152
non-racialism, women's experiences of, 152
Women's Charter, 152, 177-178
First, Ruth, 25, 65
Fischer, Bram, 3, 57, 99-101, 102, 113(n65), 151
Fischer, Molly, 151
Five Freedoms Forum (FFF), 254
campaigns, 257
origins, 257
publications, 258
Forman, Lionel, 130
Forum Club, 124, 130
Fourie, Joey, 103
Franchise Action Council (FRAC), 65
Freedom Charter, 2, 46, 127, 161, 163, 204, 215, 220
and Africanists, 163
controversy surrounding, 180-181
critics of, 181-187
drafting, 161, 173, 175-180
Million Signatures Campaign, 187-189
nationalization clause, 178-181
in the 1980s, 243, 249, 250, 251, 266
People's Charter preceding Freedom Charter, 46
ratification of, 187
significance, 192
Friends of Africa, 18
Friends of the Soviet Union, 86, 93, 103

Garment Workers Union, 107
Goldberg, Denis, 82, 226
family history, 88-89
in MK, 231-232
politicization, 89, 96-97
and Treason Trial, 191
Gomas, John, 17
Goldberg, Vic, 212, 213, 214, 222n
Goldreich, Arthur, 233-234
Gordimer, Nadine, 139-145, 154, 158
Gosschalk, Bernard, 146
Guevara, Ché, 231, 238
Gumede, J.T., 14, 16

Harmel, Michael, 25, 41, 48(n1)
election activity, 37
and Freedom Charter, 181
Johannesburg Discussion Club, 129
and whites, role of, 132

Harris, John, 237
Hashomer Hatz'air. *See* Jews, Zionist-Socialists
Hayman, Ruth, 207
Hepple, Bob, 233
Hirson, Baruch, 20n, 153(n22), 212, 214, 222n
  in NCL, 235, 236, 238, 240(n21)
Hodgson, Percy Jack
  in COD, 73
  Congress Alliance, socializing in, 149
  early history, 105
  in MK, 232
  in Springbok Legion, 57
Hodgson, Rica, 149, 232
Hogan, Barbara, 249
Holt, Margaret, 66
Home Front League. *See* Springbok Legion

Ikaka Laba Sebenzi, 86
Immigrants
  British, 3, 11, 105-111
  Jews, 3. *See also* Jews
Industrial Workers Union of Africa, 12
Institute for Democratic Alternatives, 254, 263-264
Institute of Industrial Education, 247
International Socialist League (ISL), 12, 40
  and African nationalism, 12
  and Africans, 12
  and SALP, 12
  Yiddish-speaking branch, 85
Isacowitz, Jock, 57, 77(n35), 165

Jewish Board of Deputies. *See* Jews
Jewish Workers Club, 85, 95, 124
  and Birobidzhan, 86
  and CPSA, 86
  and Soviet Union, 86
Jews
  and Afrikaners, 98
  and anti-semitism, 84, 92, 93, 98
  assimilation, 97
  of British origin, 83, 86
  Bundists, 85, 87
  in the Western Cape province (1800s), 83
  in COD, 83
  in Communist Party, 83
  conservatism, 90, 91
  early settlement, 83
  Eastern European, 83, 84
  immigrants, 3, 83, 89, 97-98
  Jewish Board of Deputies, 85, 90, 91, 98
  and National Party, 98
  occupations, 84, 87, 89
  politicization, 91-92, 93, 94, 96, 97
  population, 83-84
  population, composition of, 83
  upward mobility, 87
  voting behavior, 90
  working class, 88, 97, 99, 124
  and Zionism, 85
  Zionist-socialists, 86, 94, 95, 97
  Zionists and socialists, 85
Joffe, Max, 92
Johannesburg Democratic Action Committee (JODAC), 243, 258
  and CC, 257
  constituency, 255
  and FFF, 257
  and liberals, 255, 258
  non-racial contact, 258
  origins, 254-255
  self-criticism, 255
  and townships, 256
  white community, work in, 257
Johannesburg Discussion Club, 124
  and Congress Alliance, 126, 128, 129

and indigenous bourgeoisie, 127-128, 129
and national-democracy, 127
National Question, 125
origins, 135(n22)
society, class nature of, 125-126, 127
and Soviet theorists, 126, 127
and working class, 128
See also discussion clubs
Joint Council Movement, 18, 100
Jones, David Ivon, 12, 105
Jordaan, K.A., 131-132
Joseph, Helen, 106-108, 243
in AES, 106
early history, 106
in FSAW, 152, 153-154
politicization, 106
Joseph, Paul, 150-151
Junod, Violaine, 165

Kahn, Sam, 17, 37, 40-42, 119
Kane-Berman, John, 77(n35)
Kasrils, Ronnie, 233
Kathrada, Ahmed, 233
Kemp, Stephanie, 238
Kodesh, Wolf, 170
in COD, 147, 156
early life, 92
family history, 84, 87-88
in Springbok Legion, 58-59
Kotane, Moses, 17, 26, 31, 43, 150, 191

Labour Party. *See* SALP
La Guma, James, 14, 15, 16, 17, 43
League Against Imperialism, 14
League of African Rights, 16
Leballo, Potlako, 184, 186, 200
Lee-Warden, Len, 73
Left Book Club, 86, 93, 124
Leftwich, Adrian, 235, 236, 237
Lembede, Anton, 5, 19, 29, 70
early history, 27-28
and liberalism, 29
and national convention idea, 46
Letele, Arthur, 180
Levy, Leon, 173
Levy, Norman, 97
Lewin, Hugh, 236, 237, 238
Liberalism, 6-7, 28
interwar, 18-20
Liberal Party, 57, 201, 214, 219, 235
in African areas, 210
and ANC, 165, 207, 208, 209
and broad front, 206, 211
and COD, 8, 164-165, 167, 202, 206, 207, 208, 209, 211
and communism, 8
and Congress Alliance, 8, 192, 207, 208, 209, 210, 211
and Congress of the People, 162, 165-166, 175: as internally divisive, 165-167
and extraparliamentarism, 205, 206
formation, 7, 71
and Freedom Charter, 180
legalism of, 210
and Multi-Racial conferences (1957, 1958), 208
non-racialism of, 7, 8, 211
radicalization of, 197
and State of Emergency (1960), 228
and Treason Trial, 191
Liberals
and COD formation, 69, 115, 146
and SACP, 116
Lipman, Alan, 90
London Missionary Society, 7
Lutuli, Albert, 147, 174, 180, 189, 191, 198, 199, 200

Madzunya, Josias, 186
Makabeni, Gana, 103

Makiwane, Tennyson, 228
Malan, Daniel F., 46
Mandela, Nelson, 28, 156, 181, 186, 200, 232, 266
Marks, J.B., 25, 86, 150
Marquard, Leo, 199
Mashaba, Bertha, 152, 153
Matshikiza, Todd, 148
Matthews, Joe, 119, 145
Matthews, Z.K., 161, 166
Mbeki, Govan, 233
Mda, A.P., 28, 70
Mhlaba, Raymond, 47, 233
Million Signatures Campaign. *See* Freedom Charter
Mntwana, Ida, 152
Modern World Society, 124
Modern Youth Society, 97, 124
Modise, Joe, 233
Mofutsanyana, Edwin, 25, 46, 86
Molteno, Donald, 43
Moosa, Rahima, 152
Moroka, James, 47
Mothopeng, Zephania, 200
Mphahlele, Ezekiel, 173-174
Mtini, John, 170
Mtolo, Bruno, 234
Multi-Racial conferences, 197-201, 208, 211
 ANC, role in, 198, 199-200
 IDAMF, role of, 198, 200

Naicker, Monty, 44, 66
Naidoo, H.A., 43, 104
Natal Indian Congress, 47
National Committee for Liberation
 activity, 235, 236
 aims, 237
 composition, 235, 240(n21), 240(n22)
 formation, 230
 and Liberal Party, 226, 236
 structure, 235
 theory guiding, 238
National convention idea, 42, 45-46, 1, 162, 200, 230, 239, 240(n10)
National Liberation League (NLL), 17, 40, 43
National Union of South African Students (NUSAS), 236, 246, 248, 249, 250, 257
Naudé, Beyers, 249
Ndaba, Hezekiel, 182-183
Ngoyi, Lillian, 103, 152
Ngubane, Jordan, 28, 180, 208
Ngwevela, Johnson, 25
Nkosi, Lewis, 148-149, 151
Nokwe, Duma, 200, 216
Non-European United Front, 17
Non-European Unity Conference, 43
Non-European Unity Movement (NEUM), 119
Non-racialism, 1, 2, 108, 239
 achieving, process of, 226, 265-266
 barriers to, 215
 conceptualizing, 264-265
 cultural dimensions of, 174-175
 durability of, 4
 experiences of, 199, 213, 262-263, 264, 265
 and Freedom Charter, 244
 in the 1980s, 243, 244, 250, 266-268
 socializing, 139-145, 148-152: complications of, 151-152
 in trade unions, 247, 264
See also ANC, One Congress question *and* COD, One Congress question
Nzula, Albert, 16

Odendaal, André, 262-263
O'Meara, Rhona, 35

Palmer, Josie, 25

Pan-Africanist Congress (PAC), 33, 119, 240(n10)
and ANC, 6
formation of, 6
and Sharpeville massacre, 227
violent activity, 230-231
Paton, Alan, 205, 208
People's Assembly for Votes for All, 46
People's Cooperative Trading Society, 104
Picardie, Michael
early history, 95
family history, 84
and Million Signatures Campaign, 188, 189
politicization, 96
in State of Emergency (1960), 228
township experiences, 148
Potekhin, I. I., 130
Prager, Fred, 236, 238
Programme of Action, *See* ANCYL
Progressive Federal Party (PFP), 255, 257, 259
Progressive Party, 205, 210

Radford, Betty, 49(n26)
Reeves, Ambrose, 199, 201
Reitstein, Amy, 147
Rheinalt Jones, J. D., 18, 19
Richter, Maurice, 17
Rivonia Trial, 181, 233-234, 236
Rodda, Peter, 206, 215
Roussos, Mike, 264-265
Roux, Edward, 13, 15, 16, 17, 102, 113(n71), 132, 136n
Rowntree, boycott of, 250

Sachs, Albie, 238, 239
in Defiance Campaign, 66, 67
in MK, 232
Sachs, E.S., 102, 103, 107, 113(n70), 232
Sampson, Anthony, 158
Sampson, H.W., 11
Segal, Ronald, 212, 213, 214, 222n
Seme, Pixley ka Izaka, 28
September, Reginald, 170
Shall, Sid, 66
Sharpeville massacre, 197, 225-229
Simons, Jack, 15, 17, 25, 48(n1), 200
and Freedom Charter, 181
theorizing, 130-131
Sisulu, Walter, 28, 29, 68, 69, 156, 184, 233
Slovo, Joe, 4, 25
and the Congress of the People, 166, 172-173
in MK, 232-233
Smuts, J. C., 13, 54, 62
Snitcher, Harry, 25, 37
Sobukwe, Robert, 200
Social Democratic Federation, 11
Socialism, 11
Socialist League, 235, 240(n21)
South Africa Club, 124, 130
South African Bureau of Racial Affairs (SABRA), 200
South African Coloured People's Organisation (SACPO), 74, 119, 245
and Congress of the People, 170
South African Communist Party (SACP), 4, 5 73, 234, 248, 266
and COD, 134n
"Colonialism of a Special Type" thesis, 33, 126, 127, 231
and Congress Alliance, 126, 128
and MK, 225
and sabotage, 230
South African Congress of Democrats. *See* COD
South African Congress of Trade Unions (SACTU), 179, 247
and State of Emergency (1960), 229

South African Indian Congress (SAIC), 119
and ANC, 44, 65, 163, 245
and Congress of the People, 167
and FRAC, 65
and State of Emergency, 229
South African Institute of Race Relations (SAIRR), 7, 18, 19
in the 1980s, 254
South African Jewish Board of Deputies. *See* Jews, Jewish Board of Deputies
South African Labour Party (SALP), 11, 88, 207
and ISL, 12
and Labour League of Youth, 94
in Pact government, 18
and WWI, 12
South African Liberal Association. *See* Liberal Party
South African Students Organisation (SASO), 246
South African Typographical Union, 11
Springbok Legion, 71, 75, 95
and AES, 58
and Army weapons policy, 56
anti-fascism, 58
blacks in, 55-56, 60
branches, 56
and British Empire Service League (BESL), 57
and COD, 53, 72, 74
and Congress Alliance, 74
and constitutionalism, 61
and CPSA, 57
critique as radical organization, 53
and Defiance Campaign, 68
and demobilization, 57
dissolution, 74
formation, 55, 75n
Home Front League, 56, 57
internal divisions, 61
internal transformation, 60
leadership, 57
and Liberal Party, 72
membership, 56
and Nationalists, 61
and national liberation movement, 64
Non-European Soldiers' Dependants' League, 57
post-war activity, 60
radicalization of, 64
Services Security Code, 57
Springbok Legion Soldier's Manifesto, 55
and Suppression of Communism Act, 61
and Torch Commando, 61, 63, 64: critique of, 63
and United Party, 63, 64
See also Torch Commando
Stamelman, Selma, 66, 67
Stewart, Robert, 11
Students' Liberal Association, 96
Students' Radical Movement, 248
Suttner, Raymond, 1, 2, 253

Tamana, Dora, 146
Tambo, Oliver, 68, 69, 166
Thornton, Amy, 89, 97, 169, 170
Tloome, Dan, 150, 184, 185
Torch Commando, 65, 68, 107
leadership of, 77(n31)
and National Party, 77(n29)
origins, 62, 77(n28)
Steel Commando drive, 62, 77(n34), 77(n35)
United Party, relations with, 62
Transvaal Indian Congress (TIC), 45, 243
Transvaal Indian Youth Congress, (TIYC), 162, 211
Treason Trial, 100, 189-192, 197, 208, 219, 243

Congress Alliance, uniting, 163, 190, 191
non-racialism, promoting in Congress Alliance, 157-158, 189-190
Treason Trial Defence Fund, 191
Trew, Tony, 238
Troup, Freda, 66, 67
Tshunungwa, T. E., 169-170, 172
Turner, Richard, 246
Turok, Ben, 206, 219
and Congress Alliance, 146
and Congress of the People, 170, 173
and Freedom Charter, 179-180

Umkhonto we Sizwe (MK), 82, 236, 245
activity, 233
aims, 234
and Congress Alliance, 231
formation, 225-226, 231
Operation Mayibuye, 231, 233, 240(n13)
technical preparations, 232-233
theoretical underpinnings, 231
training, 232
whites, role in, 225, 231-233
Ungar, André, 91
United Democratic Front (UDF), 1, 108, 244, 254, 255, 260, 263
Call to Whites Campaign, 256
Congress Alliance, comparison, 251
Declaration, 251
formation, 243
origins, 251
State of Emergency (1985), effect of, 257-258
whites in, 251
See also Area Committees
United Party, 40, 62, 211
and Jews, 90
and National Party, 64
and State of Emergency, 229
Unterhalter, Jack, 167, 209, 210

Verwoerd, Hendrik, 28, 100
Vigne, Randolph, 235
Vine, Owen, 206

Wages Commissions, 246, 247
War on War League, 12
War Veterans Action Committee, 62
Watson, Robert, 235
Watts, Hilda, 25, 41
Congress Alliance, socializing in, 149-150
Electoral activity, 37-40, 50(n42)
in FSAW, 152
in Labour League of Youth, 94
Weinbren, Ben, 16, 50(n26)
Whites
acculturation, 26
activism in 1970s, 246-251
and African nationalism, 216
ANC, identifying with, 110, 146
as anomalous, 70, 71, 81, 115, 155
in black areas, 153, 169, 256, 264
blacks, relations with, 26, 95, 96, 97, 108
as domineering, 108
as elitists, 124
material advantages, 3, 5, 26, 155-156, 172
motivations, 140-145, 239
politicization, 89, 90, 91-92, 93, 96, 100, 106, 107, 168, 248, 263
prison experiences, 67, 82, 249
role in 1980s, 252
skills of, 231-233
as theorists, 5, 116, 123
theorizing, 1980s, 252, 253
uniqueness, sense of, 219, 253
working class, 105

Williams, Cecil, 57
Wolfson, Issy, 25
Wolheim, Oscar, 166
Wolpe, Harold
  family history, 94
  Rivonia Trial, 234
Wolton, Douglas, 15, 16, 17
Wolton, Molly, 17
Women's Auxiliary Air Force
    (WAAF), 54, 106
Women's Charter. *See* FSAW
World War II, 54, 89
  Africans, influence on, 54
  and CPSA, 54
  and liberalism, 54

Xuma, Alfred B., 44, 46

Yengwa, M.B., 171

Zionism. *See* Jews
Zionist-socialists. *See* Jews

About the Book and Author

Although apartheid in South Africa has generally pitted the oppressed black majority against the powerful white minority, throughout much of the twentieth century, small numbers of white South Africans have chosen to identify with African aspirations and to work for the end of white supremacist government. In this book, Joshua N. Lazerson examines the role played by self-described "white democrats" in the key formative years of the national liberation movement led by the African National Congress.

Lazerson provides a collective biographical analysis of these white South Africans, the varied paths that led them to political activism, and their motivations in breaking from white society. He also explores the impact of the white resistance on the daily workings and ideological directions of the national liberation movement.

In Lazerson's view, the white presence made it more difficult to define the nature of the struggle against white domination, because it appeared to contradict the common definition of the anti-apartheid struggle as a strictly black-against-white conflict. He explores the conflicts engendered by whites playing important, even prominent, roles in the major anti-apartheid initiatives, giving particular attention to the white presence in the Communist Party of South Africa, the Springbok Legion, the South African Congress of Democrats, and Umkhonto we Sizwe. In his final chapter, Lazerson discusses white participation in the renaissance of mass anti-apartheid activity in the 1980s and assesses the outlook for attaining a truly non-racial post-apartheid society in South Africa.

Joshua N. Lazerson, who holds a doctorate in African history from Northwestern University, is a program coordinator at the Urban Corps of San Diego.